QUANTUM FAITH

STEVE MCVEY

TWS | THE WRITER'S SOCIETY PUBLISHING

Paperback: ISBN 978-1-961180-03-1

TWS Publishing
Lodi, CA
www.thewriterssociety.online

Contents

Introduction

If someone had told people just one generation ago about the things we could do today, they wouldn't have believed it. We carry a device in our pocket that can not only make phone calls but, through the internet, instantly connect to the totality of world knowledge, take pictures and videos, send messages, and run apps that people would have considered science fiction then.

In one short generation, technological advances have handed us a list of seemingly endless abilities. Through social media, we can connect with people worldwide and immediately share information and content that would have been beyond imagination then. Cars are on the road now that drive themselves without human involvement. Machines do jobs and make decisions on their own without human direction. Biotechnology has given us the ability to modify biological systems that include DNA and genes so that diseases can be cured, and human abilities enhanced. There are 3D printers that almost seem to create objects magically. Everyday devices in our homes can communicate with each other and with us so that we can turn off our TV, turn on our lights, or change the thermostat just by talking to it. We can ask our devices from Amazon, Google, or Apple to play music, give us a weather report, remind us of something later, give us driving directions or hundreds of other things. Artificial intelligence is emerging with such speed and strength that Sundar Pichai, CEO of Google, recently said its impact is more profound than the discovery of fire or electricity. [1]

The progress leaves us almost breathless when we stop and look at it all. With the increasing advances, we are constantly pushing the boundaries of what is possible and transforming our world. As we progress on this trajectory, the potential for even greater advancements is limitless. Science has the power to change the world for the better, and the possibilities are mind-bending.

Then there is faith. Where does that fit into all this? Are science and faith incompatible? The two have often been seen as being in conflict with each other, and, unfortunately, the sort of science we were taught in school frequently stood in sharp contradiction to what we valued as biblical truth. As a result, believers had to choose between their faith and what they were taught in science class.

> Things are changing now. People today are beginning to recognize that modern science can complement faith and that a deeper understanding of both can lead to a richer and more meaningful life.

That's because science has, in one sense of the word, converted. The scientific materialism of yesteryear has given way to a new understanding that not only acknowledges but affirms many of the things we learned from the Bible as children. German physicist and an early pioneer of quantum mechanics, Werner Heisenberg said, "The first gulp from the glass of natural sciences will turn you into an atheist, but at the bottom of the glass God is waiting for you. [2]

At the beginning of the 20th century, quantum science began to emerge, gaining momentum and credibility among even the fiercest skeptics. It has upended some of the teachings of old-school science that reduce reality to nothing more than physicality. Increasingly, scientists recognize a metaphysical world they've not previously acknowledged.

Physicist and mathematician Freeman Dyson said, "The more I examine the universe and study the details of its architecture, the more evidence I find that the universe in some sense must have known we were coming." [3] So here is another scientist suggesting that there may be a more profound, underlying purpose to the universe beyond what can be explained by purely empirical evidence and mathematical means.

John Wheeler, widely recognized for his groundbreaking work in quantum theory, wrote, "As we survey all the evidence, the thought insistently arises that some supernatural agency - or, rather, Agency - must be involved. Is the present universe the product of chance or design? Was it created or merely revealed? Who created it, and how? When? Why?" [4] Unlike many of his predecessors, Wheeler raised questions about the universe's origin, the role of chance and design, and the existence of a creator. Science has come a long way!

While astronomer and cosmologist, Carl Sagan, was an avowed agnostic, he had the intellectual honesty to say, "Science is not only compatible with spirituality; it is a profound source of spirituality. When we recognize our place in an immensity of light-years and in the passage of ages, when we grasp the intricacy, beauty, and subtlety of life, then that soaring feeling, that sense of elation and humility combined, is surely spiritual." [5]

By exploring the interconnections between modern science and spirituality, we can better understand the world and our place in it. Science is now teaching us things about the culture of the kingdom of God that many are finding to be very helpful.

> I want you to see that we don't have to be suspicious about connecting scientific truths to biblical teaching, as if it somehow diminishes what the Bible says for us to allow science to inform our understanding. God is the God of all things, including true science. In the early days of science, Isaac Newton said, "The more I study science, the more I believe in God." True science is a friend of faith.

There have always been things the Bible instructs us to do that may not have made sense until science explained it. The Bible tells us what we are to do. Now science often tells us why we have been told to do that.

For instance, the Old Testament contains commands regarding health that have now been shown to have a scientific basis. For example, many Old Testament laws, such as those regarding hand washing and food preparation, are now recognized as essential measures to prevent the spread of germs and diseases.

Sexual health is another example. The Old Testament explicitly prohibits sexual promiscuity, which has now been shown to be associated with higher rates of sexually transmitted diseases. Yet even before they knew why God gave the prohibition, He knew and was protecting those He loved.

There were also the Old Testament dietary laws, such as the ban on consuming pork and shellfish, which are now understood to have been based on practical health concerns. Before modern sanitary precautions, pork and shellfish were often likely to contain harmful parasites and toxins that can make people sick when eaten. So, again, the people were being protected even when they didn't understand why the prohibition was in place.

Another Old Testament command regarding quarantine for individuals with infectious diseases, such as leprosy, is similar to modern public health measures to prevent the spread of contagious diseases. Social distancing for health reasons is as old as the Bible. Now we know the prohibition was rooted in medical science.

Not knowing the science behind the instructions, Israel assumed these commands were to conform to a moral code God gave, but today we know why God gave them these commands. They couldn't understand the science of it, but He did.

The same is true today. There are things the New Testament has told us regarding what we are to do, but now science is explaining those things' why and how.

For instance, Jesus told His disciples to "have faith in God." After cursing a tree that bore no fruit, the disciples saw later that day that the tree had shriveled up and died. Mark 11 describes the scene:

"Peter remembered it and said, 'Look, Rabbi! The fig tree You cursed has withered.' 'Have faith in God,' Jesus said to them. 'Truly I tell you that if anyone says to this mountain, "Be lifted up and thrown into the sea," and has no doubt in his heart but believes that it will happen, it will be done for him.'"

Jesus had just performed a quantum miracle that was sourced in focused intention and quantum resonance, but it looked like magic to the disciples. It wasn't. He explained to them that they too could do these sorts of things by faith, but He didn't tell them how to practice that kind of faith. Why didn't He explain how? Because as John 16:12 says in the *Aramaic Bible in Plain English*, "I have much to say to you, but you are not able to grasp it now." They couldn't grasp it so there was no need for Him to explain it. Today, two millennia later, you can grasp it, and knowing the "how" behind the "what," can change your life.

In the following chapters, we will discuss five scientific realities that will unpack and explain biblical faith to you in a way you have probably never seen before. We will combine science with biblical teaching. As you put into practice the things you'll learn here, you will become equipped to come to the place where things around you change. This isn't a magic potion, mystical mantra, or set of religious rules that will give you absolute control over everything. Only God possesses that ability. What it will do is equip you to have a greater influence in outcomes in your life than you have experienced until now.

Chapter 1

Every possible outcome already exists.

Faith is the substance of things hoped for, the evidence of things not seen.

– Hebrews 11:1

The unseen world inside the Kingdom of God is full of possibilities, just waiting to manifest in our personal experiences. This realm of divine love existing beyond our physical senses is a place of infinite potential, where everything from our deepest God-given desires to our highest dreams can come to life. By tapping into the flow of the Spirit who governs the unseen world, we can unleash the full potential of that world and co-create, together with Him, the life we were made to know and enjoy.

. . .

Ultimate Reality is defined by love. Augustine, one of the greatest Christian thinkers of all time, affirmed, "God's love is the glue that holds all things together, the very fabric of the universe." To look into the transcendent world within which this temporal world exists is to see a place where you are eternally endeared. If Augustine was correct, the universe is tilted in your favor, or, to echo the psalmist, "This I know, God is for me" (Psalm 56:9).

To glimpse into that world is to see an invisible Kingdom of Kindness where there are no limits to what you can receive and achieve by grace. It is the point where you can connect with the One who will guide you on your journey in this visible world. Whether you want spiritual growth, deeper relationships, or greater progress in other areas of life, the Kingdom of God is a Field of Favor that holds the key to unlocking your true potential and making your God-given dreams a reality.

> People often think that faith creates something out of nothing. It doesn't. Faith causes things that already exist to manifest in spacetime. Hebrews 11:1 says that faith is "substance." "Sub" suggests it takes a "stance" on an underlying reality that is already there. Faith is "evidence" of things not seen but which do exist. Everything necessary to give you the life experience you were born to know is there, waiting for you.

. . .

In the 1920s, the Danish physicist Neils Bohr and his colleagues debated the nature of reality when they agreed that all possibilities exist inside the quantum world in a state called "superposition." Superposition suggests that a quantum particle, such as an electron or photon, can exist simultaneously in multiple states or positions. A weird thing about this quantum reality is that something can exist in more than one way and more than one place at the same time.

The Copenhagen interpretation of quantum mechanics is one way to understand the strange rules of the quantum world. It says that tiny particles like electrons exist in a blurry state of probability until we measure or observe them, at which point they pop into a definite place. This "observer effect" in science says we influence manifestation by focusing on something. Until it materializes in our experience, the outcome is simply a possibility in another dimension.

Before it manifests, a quantum particle is described by a wave function, which represents the probabilities of finding the particle in different positions (possible outcomes). This wave function is a superposition of all possible states, meaning the particle is simultaneously in all of these states.

. . .

When an observation is made, the wave function collapses into a single definite state, and the particle comes to one specific position. This process, known as a wave function collapse, is a central element of the Copenhagen interpretation of quantum mechanics. Another way this collapse is often described is "popping a quiff," which refers to the "quantum wave function" manifesting as one identifiable outcome.

In essence, superposition and wave function collapse implies that the quantum world contains all possible outcomes until observed, at which point the wave function collapses and one definite outcome manifests. It is to your benefit to understand this point: Every outcome is already there! Faith doesn't create anything but calls forth a manifestation that shifts its location from the invisible to the visible. Faith moves it from potentiality to actuality.

> Jesus said that "all things are possible to the one who believes." The heart of the matter isn't whether the outcome is possible. It is. The crux of the issue revolves around our understanding and practice of faith. Faith is not wishing and hoping. It knows without a doubt.

Faith is the answer to the prayer, "Thy kingdom come. Thy will be done on earth as it is in heaven." Every detail of God's Kingdom already exists and is available to manifest in this world when we approach it through faith.

Faith doesn't cause something to come out of nowhere but causes it to come from Somewhere! When you see a dish on the menu in a restaurant and then order that dish, you expect that item to come to your table. Your attention is on it, and you are already imagining how much you will enjoy eating that meal, and, in time, it usually happens. (The quantum world is a world of probability, not absolute control.) When the meal comes to you, did you create it? Did it come from nowhere? No, it came from the kitchen and manifested at your table.

That's how it is with the Kingdom of God. Every "dish" (outcome) is already there inside the Divine Matrix. You don't create it, but instead, you facilitate it coming to you by faith. Sometimes you don't get what you want, but you will always be served the best meal for you. (In other words, if the chef is your wife and you have diabetes, she may try to steer you away from a big piece of pie every day with dinner, but she'll still see to it that you eat well.)

> The deficit we perceive in our lives is not because the answer to our prayer isn't there. The problem is in our limiting beliefs. I challenge you to open your mind to the reality that nothing is impossible with God and that all

> things are indeed possible to the one who believes.

LIMITING BELIEFS

Limiting beliefs are negative thoughts or ideas about ourselves and life itself. They can be based on past experiences, societal norms, or other factors influencing us. They can be incredibly dangerous when they prevent us from realizing our full potential and limit our ability to experience the things intended for us. Many scientific studies have proven how debilitating limiting beliefs can be.

One study done by the Yale School of Public Health found that older adults who held negative beliefs about aging, such as being less capable than younger adults, had a higher risk of developing Alzheimer's disease than those who held an optimistic viewpoint on aging. [1]

Another journal, *Psychology of Sport and Exercise*, reported that research showed that athletes with limiting beliefs about their abilities paid the price for it. Those who thought they weren't as skilled as their competitors performed more poorly and were more likely to give up on challenges than those who had more positive beliefs about their abilities. [2]

Research published in *Psychosomatic Medicine* found that people who held negative beliefs about their ability to cope with stress, such as believing that they were not resilient or didn't have effective coping skills, had poorer health outcomes, including higher levels of inflammation and higher rates of chronic diseases. [3]

A study published in the journal *Personality and Social Psychology Bulletin* found that students who held fixed beliefs about intelligence, such as believing that their intelligence was a fixed trait that couldn't be improved through effort or practice, had lower academic achievement and were more likely to give up on challenging academic tasks than those who held more growth-oriented beliefs about intelligence. [4]

Limiting beliefs can negatively affect you in many ways, including the way you express your faith. Limiting beliefs are based on lies and are kryptonite to super faith. They must be dealt with if you're going to realize the things that have been divinely intended for you. Limiting beliefs have developed over a long time. Usually, they aren't instantly disarmed, but it is possible at any moment to begin to align ourselves with Ultimate Reality so that they gradually dissipate until they finally disappear.

The Example of Moses

One of the most outstanding spiritual leaders in the Bible was Moses. He is mainly remembered as the man who led the Israelites out of Egyptian slavery and as the one to whom God gave the Ten Commandments on Mount Sinai. For millennia, he has been honored as a superstar of faith. But the fact is that even a man of his stature was challenged by limiting beliefs. He might be unknown today if he hadn't found freedom from their restraints.

When God first appeared to Moses in the burning bush, Moses was overwhelmed with doubts about his abilities. He asked, "Who am I that I should go to Pharaoh and bring the Israelites out of Egypt?" He stood on the edge of an opportunity that would shape the destiny of a nation, but it all hung in the balance that would be decided by whether or not he would overcome the limiting beliefs that stirred his fears. You know that he did overcome those limiting beliefs and acted in courage, leading Israel out of slavery.

> It might be a good time for you to consider a related question: What divine plans for your life might be thwarted or even stalled by limiting beliefs that hold you back? Moses found the strength and courage he needed, and so can you.

It's important to note that finding freedom from limiting beliefs that have debilitated us for a long time isn't instantly and magically attained. Instead, growth in this area is a learning process. It was one that Moses had to have reinforced repeatedly as he led the nation of Israel through the desert toward the Promised Land. He faced situations over and over again that put him to the test.

. . .

For instance, there was the day when the people found themselves trapped between the Red Sea and Pharoah's advancing army. Thinking their destruction was imminent, they were filled with terror and desperation. They cried out to Moses, who, in turn, found a private place and expressed the same fear to God. He was still battling the limiting belief that he wasn't equipped to handle certain circumstances in which he found himself.

In His grace, God spoke to Moses and encouraged him. Consequently, Moses inched forward, step by step, determined to move through his limiting beliefs with an immature but indisputable faith that God would bless him. Having moments earlier been forced to face his own fears, He said to the cowering Israelites, "Do not be afraid. Stand firm, and you will see the deliverance the Lord will bring you today. The Egyptians you see today you will never see again. The Lord will fight for you; you need only to be still" (Exodus 14:13-14).

Then it happened just as he had said. The Lord told Moses to stretch out his hand over the Red Sea so the waters would part for the people to cross. Likely still having feelings of fear and doubt, Moses did what God said, and the miracle happened. The people crossed on dry land, and when the Egyptian army tried to follow, the waves fell back on top of them, and they drowned.

At other times during his leadership of Israel, Moses' limiting beliefs surfaced and had to be addressed. Moses was a human being just like you. Don't think that one moment of success erased a lifetime of having been programmed with limiting beliefs. The key to his success was trusting God and moving through those false viewpoints one day at a time. He didn't wait until their voice stopped speaking to him. He moved through them, refusing to acknowledge any authority in them.

> This is the way it must be with each of us. Don't wait until there is no trace of limiting beliefs inside you. Faith doesn't require that. There may never come a time when you don't occasionally hear a voice from your past whispering doubts and fears. You may hear an accusing internal murmur at times that causes you to doubt yourself. When those moments come, don't beat yourself up. You're in good company. It happens to the strongest and most influential people.

Moses is a testament to the power of identifying limiting beliefs and refusing to be controlled by them. Despite his doubts and fears, he chose not to be ruled by them. Instead, he faced them head-on, looked through them, and saw the outcome he wanted to realize. Anything might have happened since all outcomes already exist, but Moses focused on the one he wanted to see manifest. Hebrews 11:24-26 memorializes his actions: "By faith Moses, when he had grown up, refused to be known as the son of

Pharaoh's daughter. He chose to be mistreated along with the people of God rather than to enjoy the fleeting pleasures of sin. He regarded disgrace for the sake of Christ as of greater value than the treasures of Egypt because *he was looking ahead* to his reward (emphasis added)." That's what you need to do. Look ahead, beyond the beliefs that hold you back, and see the outcome you want. Then, however tentative and shaky you may feel, move forward.

The underlying quantum truth seen in how Moses responded to limiting beliefs was his recognition that outcomes are not fixed, so that they must happen in the way would seem apparent. Evidence suggested there was no hope for Israel's survival when Pharoah's army cornered them against the Red Sea. When He stepped into what looked like an impossible outcome, Moses demonstrated a quantum reality rooted in how God designed life in this world to operate.

Science has proven that we can influence the quantum fields that underlie physical reality by focusing our thoughts and intentions on a desired result. Pascual Jordan, a German physicist who was one of the founders of quantum mechanics, wrote, "Observations not only disturb what has to be measured, they produce it."[5] Where you put your attention reveals your expectation, which is another word for faith.

. . .

In the language of the Christian faith, we have authority inside the Kingdom of God to shift and shape outcomes in this world. We don't just have to wait and see what happens. We can participate in the result.

Faith is a focused intention that requires a willingness to take action in the face of uncertainty. From both a biblical and scientific perspective, an elevated expectation that the circumstances will bend in our favor is an important aspect that empowers us to act as if the outcome we want will manifest even when the surrounding external data suggests otherwise. Don't allow limiting beliefs to have the final word. Persistently cling to the fact that the outcome you want exists, and, even if it would take a miracle to manifest the outcome, you know you live in a place where miracles are always possible.

IDENTIFYING LIMITING BELIEFS

Identifying limiting beliefs can be challenging because they are often deeply ingrained in a person's thinking and behavior patterns. The first step to overcoming them is to identify what they are. What restraining thoughts have come to you during your life? They may have come from your parents, education, church, circumstances, or other significant voices that have spoken into your life.

Take your time with this part of the process. Ask the Counselor inside you to show you the ways you have been handicapped by damaging beliefs that were imposed on you. Sometimes the messages came from well-intended people, but the damage to you is no less detrimental than if it had been intentional.

Our early experiences and relationships can have a profound impact on the beliefs we form about ourselves and the world. If we grew up in a negative environment, we may have developed debilitating ideas about our potential. What happened back then may still affect you today.

The messages we receive from society, such as our church life, can also shape our beliefs and attitudes. For example, if you were taught and accepted that you aren't worthy of divine blessings or answered prayer, how could you possibly expect to have a strong faith? It won't happen until you challenge that lie and see that children are never required to deserve the favor of a good parent. Good parents love their children because they are their children! That's especially true when that parent is their Heavenly Father.

> Imagine your child saying to you, "I'm not worthy. I don't deserve your love. I don't have a right to any good thing you have to give." You would respond, "Deserve??! What are you talking about? This isn't about deserving anything. I love you, and you don't have to earn anything. You're my child!"

. . .

That kind of situation happening in a typical family would instantly be recognized as bizarre, but when it comes to being children of God, some can't see it's the same. Some people have come to believe that, in one sense, they're a better parent than God because, while the whole concept of worthiness would have no place in their relationship with their own children, they think it is somehow different in their relationship to their Heavenly Father. That's just one example of how a seriously skewed perception can go unquestioned until it is called out for the lie it is.

Many other voices might have led you to limiting beliefs. Trauma can leave deep imprints on our belief system. The media we consume can shape our beliefs, both consciously and unconsciously. Our social connections can undoubtedly play a role in shaping our beliefs. Lack of exposure to alternative perspectives can lock us down inside limiting beliefs. If we aren't exposed to a diverse range of perspectives and experiences, our assumptions may become limited to the point that we can't imagine anything different from what we know.

These are a few of the many sources from which you may have received a harmful message. No one likely meant to do anything wrong. It's even possible that you misinterpreted what they wanted to communicate. What matters now is that you recognize these limiting beliefs so that you can be free from them so that faith might ignite.

Challenging Limiting Beliefs

When you have identified the limiting beliefs, challenge the evidence that supports them. This is a critical step in dismantling them. It is important to look at what you believe and determine if there is evidence to substantiate it. The goal is to separate what is true from what you may have wrongly believed to be true. Carl Sagan said, "The greatest enemy of knowledge is not ignorance; it is the illusion of knowledge." Physicist Richard Feynman agreed: "The first principle is that you must not fool yourself, and you are the easiest person to fool." Sometimes we can be sure about something when we couldn't be more wrong.

For instance, some people who grew up in church believe they don't deserve to have their prayers answered. They were infected with a doctrinal teaching that has left them thinking they should be thankful they're even going to heaven. They believe they haven't lived a life deserving of divine blessings. Sometimes they even believe that distorted view is "humility."

Until they realize how debilitating that viewpoint is, experiencing a robust faith won't come easily. "Are you saying we deserve to go to heaven?" one might ask. The question misses the point entirely. A life in grace is not about deserving. Grace is unilateral goodness that comes without merit on the part of the recipient.

> Challenge the evidence of the limiting beliefs in your life, regardless of where they came from. This can help you to see them differently. It can be the start of a transformation that can release you from being debilitated and free you to a more wide-open space than you've ever known. It's in that wide-open space those amazing things manifest. This is a universal truth that applies to everyone.

Consider Elon Musk, entrepreneur and CEO of SpaceX and Tesla. Here's a man who refused to entertain limiting beliefs when others tried to impose them on him. Musk's goals to transform the transportation industry and eventually colonize Mars are ambitions many people initially believed bordered on the absurd. Despite facing obstacles, from technical and financial barriers to skepticism from others, he has relentlessly pushed ahead in pursuing his vision. He has repeatedly stated his belief that humanity must become a multi-planetary species to ensure our survival in the long run. While his projection of the potential future of mankind may be up for debate, no one can deny his dogged determination that evidences his strong belief in achieving tasks that most have considered impossible.

His commitment with SpaceX to develop reusable rockets that can greatly reduce the cost of space travel, led to an outcome that borders on the miraculous. In 2015, SpaceX sent a rocket into space and landed it back on earth after the mission. It was an achievement never seen before.

Musk's belief in the potential of electric cars is another example of how refusing to accept limiting beliefs can lead to amazing results. When he started Tesla, it was against a tide of opinion that the whole idea would never gain traction. That didn't dissuade him, so he went on to develop high-performance electric vehicles that could compete with gasoline-powered cars. Today, Tesla is one of the most valuable car companies in the world, whose market cap rivals industry giants like General Motors and Ford.

Who can deny that his passionate beliefs are an example of what can happen when we don't listen to the negative voices and are willing to persevere to overcome seemingly insurmountable obstacles? These things only happen once a person says an emphatic "No!" to limiting beliefs and sees the outcome clearly before it manifests.

Develop A New Perspective

Once you have identified your limiting beliefs and challenged them, developing a new perspective is important. Having a new perspective enables you to see things differently, and in doing that, you will broaden your view of the world. It will open you to possibilities you haven't considered in the past.

. . .

Many things will help you to develop a new perspective, but one way that seldom fails is to surround yourself with people who have different outlooks and are willing to challenge your beliefs. It is incredible how much influence the people we spend our time with can influence us. It's easy to be influenced by things that can even seem incidental in our interactions with other people.

In one scientific experiment, participants were shown videos of people with different emotional expressions and were told to mimic those expressions. Results showed that participants reported feeling the same emotions as the individuals in the videos, demonstrating emotional contagion through facial mimicry. Just mimicking another person's facial expression made them feel the same as that person.

In another study, participants were exposed to recordings of people with different emotional tones of voice and were asked to rate their own emotions after exposure. Results showed that participants reported feeling the same emotions as the people in the recordings, demonstrating emotional contagion through their tone of voice.

> If you want a new perspective, consider carefully who you allow to influence you. Listen to teachers who uplift you. Read books that introduce you to new ways of thinking. Find friends who nudge you upward and not pull

you down. Open yourself to new possibilities. Don't accept the notion that something isn't possible just because you can't see a clear path toward its manifestation. Great things materialize when we look beyond the material possibilities into the world where all things are possible.

Some people have been indoctrinated into a fear of failure that dooms them from the start. So, they look for guaranteed ways to ensure their desired outcome before they move forward. Consequently, many get stuck in analysis paralysis and never make progress. Refuse to allow that fear-driven line of thought to guide your direction. There will always be naysayers talking to you externally and internally. Pay no attention.

How will we ever learn if we only allow ourselves to be exposed to influences that do nothing but strengthen our existing paradigm? There is no growth without change. However scary it may feel at times, we must move outside our comfort zone if we are going to grow. Nelson Mandela once said, "It always seems impossible until it is done." Jesus said one could make a mountain throw itself into the sea if that person "does not doubt in their heart but believes that what they say will happen" (Mark 11:22-23). Align your beliefs with the world where impossibility yields to "Him-possibility" and watch what happens. Every possible outcome already exists, so embrace that fact in order to activate your faith.

Chapter 2

The Outcome You Get Is Influenced By the Outlook You Have.

It shall be done to you according to your faith.

–Jesus, in Matthew 9:29

Positive thinking is a popular field of study for good reason, and it's not just to make you feel good. It has been linked to improved heart health, better sleep, and a longer lifespan. In addition, people with a positive outlook tend to be more resilient and able to handle life's challenges more effectively. By cultivating a positive outlook, you can reap the benefits of improved physical and mental well-being, increased happiness, and a more fulfilling life. This isn't an empty self-help technique. As you will see, the importance of a positive outlook is substantiated by both science and Scripture.

. . .

The body of scientific research that highlights the value of having a positive outlook and its impact on both physical and mental health is continually increasing. Studies have shown that people with a positive outlook have more robust immune systems and lower levels of stress hormones and experience less anxiety and depression. Cellular biologist and author Dr. Bruce Lipton wrote, "The mind is the most powerful pharmacy in the world. Your thoughts can create health or disease, happiness or misery."[1] The Bible says, "A cheerful heart is good medicine, but a broken spirit saps a person's strength" (Proverbs 17:22, NLT).

Philosopher and psychologist William James said, "The greatest discovery of my generation is that a human being can alter his life by altering his attitudes." [2]

In Matthew 9:29, Jesus said that it will happen the way you think it will happen. Your outlook on your circumstances can immeasurably transform your life for the better when you practice faith grounded in knowledge of the love of God. An enlightened "uplook" will give birth to an expectant "outlook."

But there is also such a thing as negative faith, called fear. Its power is every bit as great as positive faith. When Job found himself in catastrophic circumstances, he spoke a very telling statement when he said, "What I always feared has happened to me. What I dreaded has come true. I have no peace, no quietness. I have no rest; only trouble comes" (Job 3:25-26 NLT).

. . .

What you think becomes the rails that bring what you get into your life. Romans 8:5 (English Standard Bible) says, "For those who live according to the flesh set their minds on the things of the flesh, but those who live according to the Spirit set their minds on the things of the Spirit." Think this and get this. Think that and get that. That's how the Bible says it works in everyday life.

For example, a person with a positive outlook on a job interview can confidently approach the interview, express themselves clearly, and present their skills and qualifications effectively. On the other hand, a person with a negative outlook may feel anxious and perform poorly during the interview, reducing their chances of being offered the job.

Our outlook also affects our resilience in facing challenges and setbacks. A person with a positive outlook is more likely to see challenges as opportunities for growth and to learn from their experiences. In contrast, a person with a negative outlook may become discouraged and give up when faced with obstacles.

An admirable example of this is Malala Yousafzai, a Pakistani education advocate and Nobel Prize laureate who was shot in the head by the Taliban in 2012 for advocating for girls' education in Pakistan. Yet, despite this severe attack, she survived and continued her activism, eventually becoming the youngest Nobel

Prize laureate in 2014 for her advocacy for children's rights, especially the right to education. Today, she is an international activist and speaker, continuing to raise awareness and advocate for girls' education and children's rights.

How does quantum science inform and fortify faith? You know from the previous chapter that the universe is not solid, and that particles and matter exist in a state of potential, called "superposition," until they are observed. This observer effect demonstrates the way that focused intention has a physical impact on the world around us. At the moment we focus our intention on a specific outcome or goal, we start to influence the potential state of reality. This idea suggests that our thoughts and intentions can shape and create the reality we experience.

> Thoughts and intentions are elements of faith. They can be identified by where you place your attention. Where you look reveals what you expect, and what you expect is what you will likely experience. Attention equals intention (expectation) and results in quiffs popping. So much evidence now exists to show that what we focus on often transitions into actual experience.

Lynn McTaggart is a best-selling author and journalist who has written extensively about the power of intention in shaping our reality. This researcher has demonstrated through empirical studies how a focused thought or desire can change the physical world. McTaggart emphasizes the importance of developing a clear and positive intention and of being in a relaxed, receptive state when setting an intention. She also teaches that it's imperative to believe in the power of your intention and to remain open to receiving the manifestation of your desires. McTaggart's teachings also emphasize the importance of being in a state of gratitude and love when setting an intention, since these positive emotions can boost the manifestation process.

The Intention Experiment

One experiment McTaggart conducted was the "Intention Experiment." It was done in collaboration with several universities and organizations and is often referenced as evidence of the power of intention. In this study, participants were asked to focus their intentions on a specific target (in one case, a beaker of water) while scientists monitored the target for any changes. The results showed that participants' intentions significantly impacted the target, with the beaker of water changing its structure in response to the focused intention.

The Remote Healing Experiment

Another experiment, the Remote Healing Experiment aimed to test the impact of intention on physical healing even when it was done a distance from the targeted recipient. Participants were asked to focus their intentions on a distant person for a set amount of time every day. The results showed a noteworthy improvement in the person's physical health, suggesting that distant intention can impact physical reality. In simple terms, the experiment showed that people could send healing thoughts to someone far away, and they could make him better.

A biblical example of this is found in the fourth chapter of John. In the town of Capernaum, there was a royal official whose son was very sick. The nobleman had heard about Jesus and believed that He could heal his son. He traveled to Cana, where Jesus was, to ask for His help.

When he found Jesus, the nobleman begged Him to come to Capernaum and heal his son. Jesus replied, "Unless you people see signs and wonders, you will never believe" (John 4:48). However, the nobleman persisted, saying, "Sir, come down before my child dies" (John 4:49).

Jesus then said to him, "Go, your son will live" (John 4:50). The nobleman believed Jesus and set out for Capernaum. On the way home, his servants met him and told him that his son was alive and well.

The nobleman asked them what time his son got better, and they replied, "Yesterday, at one in the afternoon, the fever left him"

(John 4:52-53). The nobleman realized that this was the exact time that Jesus had said, "Your son will live."

Here's an example of science and faith pointing to the same Ultimate Reality. Was it a quantum event or a spiritual event when Jesus healed this man's son? Yes! It was both! We must come to the place where we stop separating the spiritual from the scientific and recognize that true science allows us to "peek behind the curtain" and see *how* God works in our midst. He wants us to know so that we can more consistently and effectively live like the Kingdom Citizens we are.

Another example was the story told in Matthew 8 about a Roman centurion who also wanted his servant to be healed. The man came to Jesus and pleaded with Him to heal his servant, who was lying at home, paralyzed and suffering terribly. Jesus agreed to come and heal the man.

However, the centurion said to Jesus, "Lord, I am not worthy to have you come under my roof, but only say the word, and my servant will be healed" (Matthew 8:8). This man understood the reality science is finally explaining to us today.

Jesus was amazed by the centurion's faith and said to the crowd that had gathered, "Truly, I tell you, in no one in Israel have I found such faith" (Matthew 8:10). Then He said to the centurion,

"Go; let it be done for you according to your faith." The bible says, "And the servant was healed instantly" (Matthew 8:13).

> Can you see in these instances how science and faith point to the same reality? Were these quantum events? Of course, they were! They were also incidents empowered by the Spirit at work in and through Jesus Christ. God works through science. A quantum understanding doesn't dilute the miracle. I hope it encourages you to see that you are capable of a faith that sees miracles!

The experiments conducted by McTaggart mentioned earlier, among others, provide evidence for the power of intention and suggest that faith is connected to where we focus our attention and can impact the physical world in measurable ways. Your life can change for the better when you understand how the Spirit of Truth, who is also the Spirit of Science, will work through you when you have eyes to see the reality of these things.

Many people's experiences in life don't change because they have been wired to focus on the negative outcomes and only see life through a dark lens. It doesn't have to be that way. The Bible says it is possible to "be transformed by the renewing of your mind." How do we do that?

. . .

The answer is to change the way we think. When we change how we think, we change our focus, and when we change our focus, we change our expectation, and when we change our expectation, we change our outcomes! We are not victims. With the spiritual approach of scientific applications, we can change how we think, and our lives can change.

Neuroscience teaches us that we can alter our physical brains through neuroplasticity. Neuroplasticity refers to the brain's ability to change and rewire itself in response to experiences we have. It deals with the way the brain creates new neural connections and reorganizes existing ones in response to changes in our thinking, behavior, and circumstances.

The idea of neuroplasticity challenges the traditional view of the brain as a fixed and unchanging structure. Instead, it suggests that the brain has the ability to change and adapt throughout our lifetime. This happens through a process called "synaptic pruning." It occurs when the brain eliminates unused or redundant neural connections and synaptic growth, while at the same time forming new neural connections in response to new experiences and learning.

You can change how you think so that your outlook leads to a different outcome. The Bible calls it being "transformed through the renewing of your mind." It's not a complicated matter; it only requires cooperation with the Spirit of Truth for you to see it happen in your own life.

To make the process easy to remember, let's use the acrostic, GROW to represent ways to change your outlook:

G Stands for "Gather"

The first basic step forward in changing how you think is to gather information that will motivate a change. Many people live at a level where their beliefs and behavior are shaped by their subconscious. They mindlessly live the same lifestyle every day, doing the same things and getting the same results without even thinking about it.

The neural connections that have formed in their brains over time have trapped them in a rote routine that they never give any thought to. Nothing changes because they don't stop to consider the possibility of change. So, the first step in changing your outlook so that your faith begins to empower your lifestyle is to stop, take a look at the lifestyle you're living now, and then determine to gather knowledge about how things can be different.

Don't make this a rigorous regiment that requires you to rev up the religious rpms so that God will help you. It's not complicated. God has already helped you by allowing you to understand how your brain works, why you are where you are, and how to change it. Think for a moment about how your brain works and how you came to the place where you are today.

YOUR MENTAL GPS

Imagine the neurons in your brain firing with every thought you have. When they fire, electrical impulses shoot from one neuron to the other. Continuing to have the same or even similar thoughts causes more neurons to fire until the impulses connect them to each other through what science calls a synapse. Think of a synapse like a bridge that allows information to flow between neurons, which helps coordinate the functions of the nervous system.

A common expression says that "neurons that fire together, wire together." As those synaptic connections continue to join neurons, the pattern forms a sort of internal GPS directs your choices and actions. Envision a mental map with pins all over it and highlighted routes drawn that show the connection between the various pins. These routes become the map that outlines your lifestyle. Most of the time, we travel that route without even being consciously aware of it.

> An important step, then, is to become aware of the way you're thinking and introduce new ideas to your brain that will disrupt the subconscious, hardwired, habitual way of thinking.

> New ideas activate new neurons which can create new ways of thinking. When you create a new mental map, you move toward a new lifestyle with different outcomes.

Albert Einstein once said, "The world as we have created it is a process of our thinking. It cannot be changed without changing our thinking."[3] You need some new pins placed in your brain to change how you think, and that happens when we gain new information. An oft-repeated saying is "When we know better, we do better," and it's true. We come to know better as we increase our knowledge. "You shall *know the truth* and the truth shall set you free." The truth alone doesn't set us free. It's when we know it that results come.

Gathering new information that leads to change doesn't happen in one quick leap. Science studies have shown that there is a process that people move through when they experience lasting change. One study called "The Transtheoretical Model" explains how people progress through different stages when changing their lives. These stages take time, and there are six of them. Consider these stages and see if you can identify where you are in the process of change.

Precontemplation: In this stage, a person is not yet considering making a change and is unaware of the issue or problem they may need to address. Psychology calls it "unconscious incompetence." In the workplace, Human Resource Departments sometime call it "skill or knowledge gap analysis." The healthcare world calls it "clinical inertia." I mention these so you can see how pervasive this problem is across the board in society. The Bible calls it being "blind" or "lost." The problem is real. People don't know what they don't know. Maybe, in order for your faith to grow, you need to understand that there are areas of life where you need to first gain knowledge before your faith can grow. Perhaps now would be a good time for you to ask the Teacher who lives in you to show you ways this dilemma might be true in your own life.

Contemplation*:* Once we come to see that there is a gap between where we are and where we want to be, and we have taken the first step that will help us bridge that gap, it's time to look forward. Don't fall into the trap of wallowing in regrets about the past. You can't change that, but you can focus on the future your loving Father intends for you.

Long ago, Socrates said, "The secret of change is to focus all of your energy, not on fighting the old, but on building the new." The Apostle Paul would later say, "I do not regard myself as having taken hold of it yet; but one thing I do: forgetting what lies behind and reaching forward to what lies ahead" (Philippians 3:13 NASB).

Focusing on regrets about the past will only debili-

> tate your ability to move forward. If you want faith to grow, aim your focus in the direction that will nurture growth. Neuroplasticity will happen in the most productive way when you do that, and you will be transformed by the renewing of your mind.

Preparation: The most important preparation to take in anticipation of changing is to cultivate an enthusiastic mindset. A sense of optimism and resilience comes from fanning the flame of enthusiasm about what lies ahead. This is the starting place for change.

Consider that word for a moment. The word "enthusiasm" comes from the Greek word *enthousiasmos*, which is derived from the adjective *enthousiastes*, meaning "possessed by a god" or "inspired." In ancient Greece, enthusiasm was considered a divine or supernatural state, in which a person was believed to be possessed by a god or spirit, so they were inspired to perform great deeds or achieve extraordinary feats. You certainly have every reason to launch into action from a place of enthusiasm about the change to come because the Christ who created all things resides in you!

Action: Change happens when knowledge we have gained finds expression in what we do. We know, then we do. "Faith without works is dead" (James 2:26). Grace is divine enablement to rise up and live the life we have been created to know and enjoy. Don't wait until every question has been answered and every step forward has been identified. Martin Luther King, Jr. once said, "Faith is taking the first step even when you don't see the whole staircase." [4] God once told the prophet Isaiah, "And your ears shall hear a word behind you, saying, 'This is the way, walk in it,' when you turn to the right or when you turn to the left" (Isaiah 30:21, ESV). It happens one step at a time.

Quantum science suggests that your life is connected to everything around you. Physicist Fritjof Capra wrote, "The universe is not a collection of objects, but is an inseparable web of vibrating energy patterns in which no one component has reality independently from the entirety. Included in the entirety is the observer." [5]

The Bible explains, "For all things in heaven and on earth were created by him – all things, whether visible or invisible, whether thrones or dominions, whether principalities or powers – all things were created through him and for him. He himself is before all things and all things are held together in him" (Colossians 1:16-17 NET).

"All things are held together *in* Him," the Bible tells us. You are in Him and so are your surrounding circumstances. What you do affects the world around, so recognize the Divine Power in you and do something constructive to change your internal and external world.

. . .

Maintenance: Once you've started your new way of living, don't fall back. Remember that the synaptic paths need to be reinforced and the more you reinforce them, the easier it becomes to live a more productive lifestyle. This is the transformation that comes by the renewing of your mind. This is neuroplasticity. This is the way you were meant to live – at a higher level. It's time now to "set your affection (attention) on things above" (Colossians 3:2).

Don't fall back into the old ways. One biblical proverb gives a glaring example of what it looks like when people fall back: "Of them the proverbs are true: 'A dog returns to its vomit,' and 'A sow that is washed returns to her wallowing in the mud.'" When the Scripture gives a gross warning like that, it's because we're intended to pay careful attention. Don't be that guy. Maintain your walk so that your change can lead to a greater faith.

Termination: This is the level that those who understand the levels of competency call "unconscious competence." It's the stage when a person has developed a skill or knowledge in a particular area to the point that it becomes automatic, and they can perform the task without conscious effort or thought.

The Bible calls it "walking in the Spirit," because it refers to a lifestyle animated by the Spirit of Truth who indwells you. Don't think of the word "Termination" as meaning an actual finish line.

We will forever grow in our journey in faith. For our purposes, consider it reaching that place where we discover what Jesus meant when He said, "Take my yoke upon you, for my yoke is easy and my burden is light." This is the quantum life, the grace walk, or to use the Apostle Paul's words, life "in Christ."

> Don't make the mistake of thinking I am reducing this spiritual work of Spirit to a simple formula. Let's be clear: it is the Divine Life within you that will facilitate this change in you, but, on the other hand, you will be being obedient and cooperating in the process. 1 Corinthians 3:9 says, "We are laborers together with God."

These stages aren't linear, and a person might go back and forth between stages as they move through the changes they want in life. This kind of growth requires commitment, time, and trust in the One who guides the process.

R Stands for "Reflect"

There is a direct question that you must reflect on and answer if real change is going to come: Do you want to change? The question may seem silly at first glance, but it's a serious question that needs sincere reflection.

. . .

Jesus asked one paralyzed man, "Do you want to be well?" Why wouldn't the man want to be well? The fact is that, while getting well brings rewards, it also brings responsibility. It takes away excuses. It destroys any traces of a victim identity. It brings opportunity to rise up and take charge of things previously believed to be beyond our control. The bottom line is that getting well can be scary.

> Think about it. Do you want to change? Are you ready to wake up from a subconscious slumber and embrace the Divine Ability that resides within you? "Without change, something sleeps inside us and seldom awakens. The sleeper must awaken." [6] "Awake, thou that sleepest, and Christ will give thee light," the Apostle Paul wrote to the believers at Ephesus.

Reflect on whether or not you want to change. The fact that you're reading this book would suggest that you do, and that fact will give you a clear purpose and direction. It will also provide motivation to move forward toward a new outlook that can bolster your faith and lead to a new outcome.

O Stands for "Open"

Open yourself to new people, experiences, and ideas. When Joshua was about to lead the people of Israel from the familiar place they had been to a new and unfamiliar place that was much better, he said to them, "You have not passed this way before." Being open to thinking differently is an important aspect of promoting synaptic plasticity and creating new synapses in the brain.

Isaiah 43:18-19 says, "Look, I am doing a new thing! Now it springs up; do you not perceive it? I am making a way!" Don't be afraid of the new. It's there that potential resides. Old synaptic paths wither when we open ourselves to the new and embrace what is unfamiliar to us.

W Stands for "Write"

Writing in a journal can be a powerful tool for personal growth, self-reflection, and emotional healing. By putting your thoughts and experiences down on paper, you can gain clarity, perspective, and insight into your life. It is a simple but practical and powerful way to change your outlook in several ways.

First, it can be a helpful tool for tracking your progress. By regularly writing down your thoughts, feelings, and experiences, you can create a record of your personal growth and development over time. By reflecting on past entries, you can see how your thinking and attitudes have changed and gain insight into your own patterns of behavior and emotional responses.

Another way writing can help is that it focuses your intention. It helps to clarify your thoughts and make them more concrete, making it easier to focus on what we want to see manifest.

Journaling can also increase self-accountability by providing a space for honest self-reflection and self-assessment. When we record our experiences in writing, we are forced to confront our thoughts, feelings, and behaviors in a way that is often more direct and honest than we might be in our daily lives. Maya Angelou said, "Keeping a journal of what's going on in your life is a great way to help you reflect on your experiences and gain perspective on your journey."

The habit of writing your thoughts can increase self-accountability by establishing a regular practice of self-reflection and self-awareness. When we make journaling a regular habit, we are more likely to pay attention to our daily lives, and to be more intentional and deliberate in our choices and actions.

This practice can also boost confidence by providing a safe and private space to reflect on your accomplishments, strengths, and positive qualities. When we write in a journal, we have the opportunity to acknowledge and celebrate our successes, both big and small, and to reflect on the positive impact we have had on our lives and the lives of others.

By recording our achievements and positive experiences, we can build a sense of pride and self-worth, and gain confidence in our abilities and capabilities. By reflecting on our personal values and beliefs, we can become more self-aware and develop a stronger sense of identity and purpose.

In addition, journaling can help us reframe negative experiences and develop a more positive outlook on life. By writing about difficult situations and exploring our feelings and reactions to them, we can gain insight into our thought patterns and develop strategies for coping and problem-solving. By focusing on the positive aspects of our experiences and finding gratitude in even the smallest things, we can cultivate a more optimistic and resilient mindset.

Don't fall back and don't give up in your progress. Geneticist Francis Collins rightly said, "God never gives up on us, so we should never give up on ourselves or others." [7]

The outcome you get is going to be determined by the outlook you have, so be sure, before moving further into this process, to commit yourself toward the growth process necessary to establish a healthy outlook.

Chapter 3

Seeing the Outcome as a Present Reality Activates Its Manifestation.

I tell you, you can pray for anything, and if you believe that you've received it, it will be yours.

–Jesus, in Mark 11:24

Have you ever felt like the life you desire is just out of reach? No matter how hard you try, you can't seem to make your dreams a reality? If so, you're not alone. Many of us struggle to see our aspirations become tangible outcomes. But what if there was a way to make those dreams a reality? What if there was a simple yet powerful tool that could help you create the life you want? This tool is the practice of seeing what you want as a present-moment reality.

. . .

By envisioning your desires as already present in your life, you can tap into Christ's power to move into the reality He wants to give you. This practice is based on the principles of quantum mechanics, which suggest that reality is not fixed but shaped by what we believe about life. Adopting a quantum perspective on reality can align your thoughts and beliefs with biblical teaching about faith.

> When you see what you want as a present-moment reality, you focus on what you want rather than what you don't want. This helps shift your mindset from lack and limitation to an attitude of grace-filled possibility. It also helps to elevate your faith so that you can realize in your experience all the blessings awaiting you.

One key for moving to that place is to stop thinking about what's going to happen in the future and shift your focus to see that you already "have been given everything you need for life" (2 Peter 1:3). You don't lack anything but, at this very moment, have it all. Colossians 2:9-10 says that Christ is the fullness of the Godhead in human form and that you are complete in Him. It's important to denounce the idea that there is still something we need from God to give us a better future.

. . .

Inside the Kingdom of God, there is no past, present, or future. There is only the Eternal Now, and you have all you need at this moment. Stop thinking about what will happen and see it as real. God is not "going to" work in your life. The Spirit is already actively at work, guiding you into the awareness of what is already yours. It's up to you to just get up and embrace it.

This is exactly how Jesus lived. In John 5:19, He told His disciples, "Truly, truly, I say to you, the Son can do nothing of Himself, unless it is something He sees the Father doing." Jesus said that not one thing was possible for Him to do except those things which He saw His Father doing. Not what His Father was "going to do" but what He was doing in this present moment.

The problem for many who want to exercise faith is that they see manifestation in their future instead of the present moment. They say things like, "I believe that God is going to answer this prayer," "I am standing on the promises of God to do what He says," or "I know it is going to happen." That type of sentiment is often applauded in the church as "faith," but it is not the way to bring forth manifestation. Prayers and hope built on that kind of faith always look to the future, not the present.

Have you seen a greyhound run in circles around the dog track chasing the rabbit? It's called "hare coursing" and happens when a mechanical lure circles the field just far enough ahead of the dog for the animal to think that he will catch it. The fact is that the dog will never catch the rabbit because it will always be just out ahead of him.

This is the way it is with the kind of faith many people have. They believe that something is going to happen, but it never does because their faith is pointed to the future instead of the present. It's important to begin to think about the present in a different way. Using imagination is a helpful way to do that. Biblical faith is animated by a healthy imagination. Imagination allows people to visualize and create mental images of what they want so that they can see it as being already present. Don't misjudge or underestimate the power of healthy visualization. Faith is Spirit-inspired imagining of an answer to what we have wanted (prayed for).

When you use imagination as an expression of faith, you create a powerful force for manifestation and can bring your desires and beliefs into time and space. It can become real in the transcendent mind before you see it in the temporal moment. This happens because the mind doesn't differentiate between real and imagined experiences. Watch a sad movie, and you'll tear up. A scary movie will cause your body to produce adrenaline and cortisol. On the other hand, a romantic film can cause you to release "happy hormones" like endorphins, serotonin, and dopamine.

Think in a productive way, and you'll get productive results. It's all about what you see. After Lot and Abraham went their separate ways, God told Abraham in Genesis 13:15, "I am giving all this land, as far as you can see, to you and your descendants as a permanent possession" (NLT). While it's true He meant physical sight, there certainly is a metaphorical truth in that story.

God did a similar thing with another prophet; in that instance, it wasn't about physical sight at all but metaphysical vision. Jeremiah 1:11 says, "The word of the Lord came to me, saying, 'Jeremiah, what do you see?' And I said, 'I see the branch of an almond tree.'" Follow the story, and you'll learn that what he saw was a prophetic insight into what would happen in Israel.

Consider another prophet. Amos 7:8 says, "The Lord said to me, 'Amos, what do you see?' And I said, 'A plumb line.' Then the Lord said, 'Behold, I am setting a plumb line [as a standard] among My people Israel [showing the defectiveness of the nation, requiring judgment].'" There it is again. What the prophet saw was a precursor to what was going to actually happen.

> Search the phrase "what do you see?" in the Bible, and you'll discover that what a person sees is usually a prediction of what they experience. It's important, then, to learn to see. Many people need to have their vision healed to see clearly, using spiritual sight instead of physical.

If you don't see it with your eyes, then what are you to do? How do you see it? Paul wrote to the Ephesians, "I pray that the eyes of your heart may be enlightened so that you will know what is the hope of His calling, what are the riches of the glory of His inheritance in the saints." He prayed for the eyes of their hearts to see. "The eyes of your heart" doesn't refer to a physical vision but a spiritual vision consumed and carried by the Spirit who sees it all and shows you through your imagination.

Sanctified imagination is a way to see beyond the temporal and recognize what exists in the Transcendent. It's not pretending or choosing to “fake it 'til you make it.” Imagination is a power you possess that can be used for good or evil. It can be constructive or destructive. The Bible says we are to take every thought captive so that it is obedient to Christ. I encourage you to yield your imagination to the Spirit of Truth and be led into a vision of what already is when authentic faith is exercised.

Science has proven the power of imagination. One of the most widely known examples is the placebo effect. The placebo effect demonstrates the power of imagination in influencing physiological responses. When patients believe they are receiving treatment, even if it is only a placebo, their imagination can cause real, measurable physiological changes.

Another example is Mental Rehearsal. This type of experiment shows how visualization can affect physical performance. Athletes and performers have been shown to improve their abilities by mentally rehearsing their actions beforehand. See the point scored at this very moment, and every part of your body will respond to that reality.

Albert Einstein said, "Imagination is everything. It is the preview of life's coming attractions." A great tool you possess to activate faith that sees results is an imagination captured and carried by the Spirit.

Imagination in pain management is a real thing. Many studies have found that visualization can be effective in reducing chronic pain and helping people manage acute pain.

A study published in the *Journal of Pain and Symptom Management* in 2017 looked at the effects of guided imagery on pain and anxiety in cancer patients. [1] The study involved 80 cancer patients experiencing pain and anxiety related to their condition.

The patients were divided into two groups: one received standard care, while the other received standard care plus guided imagery sessions. During the guided imagery sessions, patients were instructed to envision a peaceful and calming scene, such as a beach or a forest, and to focus on relaxing their muscles and breathing deeply.

. . .

The results of the study showed that the patients who received the guided imagery sessions experienced significant reductions in pain and anxiety compared to the control group. The researchers concluded that guided imagery could help manage pain and anxiety in cancer patients.

> A healthy imagination is powerful! Don't allow yourself to be robbed of its benefit by some false teaching you may have had about the spiritual dangers inherent in the imagination. Just like your physical body, you can use it for good or evil. Begin now to use it as a tool for embracing present-moment blessings in your life. Remember, you aren't the greyhound chasing the hare that is always out in front. You are seeing the reality you want as a present-moment experience.

Amit Goswami is a respected quantum physicist who has written extensively about the relationship between consciousness and reality and has explored the idea that our thoughts and beliefs can shape our reality. He argues, from a scientific standpoint, that consciousness is the fundamental building block of the universe and that our perceptions and beliefs can influence the outcomes of quantum experiments.

. . .

He particularly emphasizes the importance of seeing what we want as a present-moment reality to have our desires come to fruition. While he is not a professing Christian, his scientific analysis aligns with the words of Jesus, who often affirmed that how we think and talk directly influences what comes into our lives.

Goswami contends that by adopting a quantum perspective on reality, we can tap into the power of our consciousness and influence the reality we want to experience. He suggests that by embracing a quantum perspective on reality, we can recognize that reality is not fixed or immutable but is shaped by our consciousness and intentions.

In particular, Goswami emphasizes the importance of seeing what we want as a present-moment reality, which can create a powerful intention to help manifest those desires into reality. By seeing our desires as already present in our lives, we can align our thoughts and beliefs with our desired outcomes and create a positive feedback loop that can help bring those outcomes into reality. Remember, this comes from a scientist who bases his views on empirical evidence.

History proves the potential of constructive imagination. You know the story of Orville and Wilbur, two bicycle mechanics who imagined flying and turned their imagination into reality by inventing the first successful airplane in 1903. You and I board jets today because of their imaginations.

Then there was Henry Ford, who imagined a world where everyone could own an automobile. The man revolutionized transportation when he mass-produced the Model T. Today, I can drive from Tampa to Atlanta in eight hours, thanks to his imagination.

NASA used their imagination to put a man on the moon, fulfilling a goal set by President John F. Kennedy in 1961. I remember watching the moon landing as a child and was amazed by how something I'd only imagined when I watched science fiction movies was now a reality because of other people's imagination.

Walt Disney used his imagination to create Disneyland, the first theme park of its kind, and later went on to build Walt Disney World, one of the most popular tourist destinations in the world. Creators there are called "Imagineers." Thanks, Walt. We've had many good times because of you.

What about the Law of Attraction?

The phrase "law of attraction" often triggers thoughts of the New Age Movement or Eastern Religions, something that scares many Evangelicals. The fact of the matter is that the origin of the Law of

Attraction is hard to pin down because it has been a concept in various spiritual and philosophical traditions for centuries. Although I suspect that many whose views differ from ours also use air conditioning, eat in restaurants, go to movies, and a hundred other things, I don't plan to give those activities up just because they do them also. I don't have to embrace everything they believe just because I utilize some of the same things.

It's true that the modern understanding of what has become known as the "Law of Attraction" as a universal principle of attracting positive experiences and outcomes through positive thoughts and emotions did become popular in the New Thought movement in the late 19th and early 20th centuries. I can't help but think of a parable Jesus told where He made the point that sometimes the children of this world are wiser than the children of Light, the idea being that some who should understand don't.

> Be careful not to throw this away just because it doesn't sound "Christian" to you. Some might find it easier to understand when it's called "the law of sowing and reaping." Don't stumble over the vocabulary.

Shortly, I will mention the difference between the Law of Attraction and what this book teaches but don't be scared off by the phrase. Truth is truth and will find its way into the world one way or the other. God is good enough that if some won't embrace the truth, others will. Divine truth is bigger than any religious viewpoint. Don't get tripped up because you've been infected with a superstitious fear wrongly believed to be some sort of spiritual check in your spirit.

I've shown you more than enough biblical evidence to see that what we think about leads to what manifests. I've described the scientific evidence for it. I've given historical examples of how imagination steered in a good direction led to the manifestation of good, God-inspired outcomes. At the risk of sounding defensive, the goal here is to quell some readers' anxiousness because of what their religious indoctrination has done to them. Not all may need this reassurance, but it's here to help those who do.

We can see things manifest in this very moment with the power of imagination. I want to review an idea from another book I wrote entitled *Quantum Life* that may be helpful:

Quantum physics comes forward to say that everything is made up of atoms, which are made of particles called quanta. The substance of those quanta is not matter but rather can best be described as energy vibrating at a certain frequency, which tends to come into existence by "gathering" with other quanta vibrating at the same frequency. The fact that, at a subatomic level, the particles that form everything are not standing still but are vibrating, translates as "everything is energy in motion."

. . .

Our thoughts are no different. They, too, are vibrating energy. Whatever thoughts we choose, they vibrate at a particular frequency that is drawn to elements (including other thoughts) that vibrate at the same frequency.

Although some have disputed its accuracy, Albert Einstein is reputed to have said, "Everything is energy, and that's all there is to it. Match the frequency of the reality you want, and you cannot help but get that reality." Einstein spoke about the relationship between matter and energy. He said, "It followed from the special theory of relativity that mass and energy are both but different manifestations of the same thing." I know you didn't get this in church, and it may be unfamiliar to you but don't dismiss this as new-age guru-talk. It is science, and it explains how the Kingdom of God works!

Einstein said that mass and energy are the same thing. Encyclopedia Britannica explains it: "$E = mc^2$ is an equation in German-born physicist Albert Einstein's theory of special relativity that expresses that mass and energy are the same physical entity and can be changed into each other."

Isaac Newton taught us that bodies with mass attract each other (called the Law of Universal Gravitation). Einstein taught us that mass and energy are equivalent through his famous formula, E =

mc^2. So, energy and matter are the same. Therefore, energy, in the form of our thoughts, attracts matter!

Do you see the simple logic of this? It is a syllogism: if "a" is true and "b" is true, then "c" must be true. It's simple logic. A. Mass attracts mass. B. Mass is energy. C. Energy attracts mass. Consequently, since internal thoughts are energy, they attract experiences in the material world. [2]

The way to interface with limitless possibilities, which science calls "potentia," and access the power of this creating principle is to focus your thoughts on the state of being surrounded by the conditions you want to produce. Set your mind and broadcast a higher vibration by thinking the right things. See the blessings you await as a finished work.

To rest in that Field of Things Above is what some call "abiding in Christ." In that environment of Pure Love, all things can happen. It's time to embrace "the mind of Christ." The Bible says, "All things are possible to him that believes," and "with God, all things are possible." Every possibility exists in wave form, in potentia, inside the Kingdom of God, where Divine Agape is the ultimate and perfect quantum life.

A critical aspect of elevating your faith to the level where it connects to and collapses the quiff you want to see pop, is your feelings. I'll talk about that in the chapter that addresses heart-brain coherence. I mention it now because I want to go ahead and note its importance, but I will explain the "how to" of it later.

In the meantime, there are a few practical techniques you can use to move toward a greater and more effective faith. Think of these techniques as the kind of exercises the Apostle Paul instructed his student, Timothy, to practice when he said, "Exercise (or some translations say "train") yourself to be godly" (1 Timothy 4:7). There is a proper place in grace to train – to exercise – so that you become godlier (and Godlike) in your life experiences.

Here are a few exercises to do every day so that you won't be trapped in a future expectation for manifestation but will come to a present experience of that manifestation.

See it

Find a quiet place where you won't be disturbed for a few minutes. Sit comfortably or lie down and close your eyes. Take a few deep breaths to relax and clear your mind. Envision yourself experiencing what you desire as if it has already happened. For example, if you want to manifest a new job, imagine yourself in your new office, interacting with colleagues, and feeling satisfied with your work. Engage your senses and focus on the details of your visualization, such as the colors, sounds, and smells. Hold the image in your mind for a few minutes and then release it, trusting it will manifest.

. . .

There's an old saying that "people are down on what they're not up on." In other words, they tend to be predisposed against things that aren't familiar. The more you practice this "sanctified imagination" method, the more comfortable you will be and the more effective it will become. It can help to set aside time each day to focus this way. Remember, this is neither a magic solution nor a legalistic formula, but a powerful tool to help you see your faith increase. With consistency and commitment, you can see it by faith and help manifest the answer you've prayed to see. You must see what God has done before you can see what God has done.

SAY IT

Speaking the right way means choosing to speak positive, encouraging, and life-giving words. When you speak words of divine love, grace, and blessing, you create an atmosphere of positivity conducive to seeing miracles manifest. By contrast, when we speak words of fear, negativity, and discouragement, we create an atmosphere that is hard to penetrate.

Speaking the right way is a simple but profound approach to changing your mind and environment. Colossians 4:6 says, "Let your speech be always with grace, seasoned with salt." That's good advice for any situation.

. . .

Research has even shown that the words we use when talking to others can have a powerful impact on their well-being. In a study published by the National Library of Medicine, researchers found that doctors who used empathetic and encouraging language when talking to patients had better patient outcomes than those who did not. [3] Words have the power to change things at this very moment!

Strengthen it

You can strengthen your present-moment experience of grace and goodness by finding the positive aspects in your life right now. One way to express that is to choose a positive affirmation that resonates with you. For example, try saying, "I am worthy because I am God's child," or "Goodness and mercy *do* follow me every day of my life!" Better yet, create an affirmation of faith specific to your needs and desires. Be sure it expresses a prayer already answered, even if it hasn't manifested yet.

Repeat the affirmation to yourself several times a day so that you can facilitate neuroplasticity and be renewed in your mind. This isn't some kind of superstition or empty approach for positive thinking. This is a proven scientific protocol. Your brain will change. Be sure to do it in the morning and before bed if you don't do it at any other time. That's the time when your mind is most open to receiving suggestions. Say the affirmation out loud or in your mind with conviction and enthusiasm. Envision yourself embodying the affirmation and feel the positive energy it creates. It may feel uncomfortable at first if you're not used to doing it, but consider it an expression of faith.

If you encounter negative self-talk or limiting beliefs, counter them with an affirmation. For example, if you hear yourself saying, "I'm not good enough," repeat, "I am worthy and deserving because of God's goodness."

Incorporating these practical applications into your daily routine can help you create an experience where the answers to your prayers manifest. Remember, the more you practice these exercises, the more natural and effortless they will become, and the more you will position yourself to see your faith grow and your prayer be answered.

Chapter 4

Understand Heart-Brain Coherence and The Door to Your Answer Will Open.

Guard your heart above all else, for it determines the course of your life.

– Proverbs 4:23

The Bible places a great emphasis on the importance of the heart, both as a catalyst for the way we behave and as a key factor in our relationship with God. Scripture teaches that the heart should be guarded because it is very important for our welfare in many ways. Most of us have believed the references to the heart are a metaphor for our inner being, including thoughts, feelings, and motivations.

. . .

Today we are beginning to realize that what the Bible says about the heart isn't only metaphorical in the way we have believed in the past. Instead, science is now showing us that the heart plays a literal role in our well-being beyond what we have imagined.

Modern science has discovered that the heart and brain are connected through a complex system of neurons, hormones, and electrical signals. The heart has its own independent nervous system known as the "intrinsic cardiac nervous system," which can communicate with the brain through various pathways, including the vagus nerve. This communication system is known as the "heart-brain axis" and is thought to play a key role in regulating emotional, cognitive, and physiological functioning.

The evidence from neuroscience suggests that thoughts and emotions can influence our experiences through their impact on neural processes and brain chemistry. Studies have shown that thoughts and feelings can activate specific regions in the brain and cause the release of neurochemicals that affect our mood and behavior.

For example, positive thoughts and emotions are associated with the release of dopamine and serotonin which promote feelings of well-being and happiness. In contrast, negative thoughts and emotions are associated with the release of stress hormones, such as cortisol, which can lead to feelings of anxiety and depression. Neuroscience research has shown that thoughts and emotions can shape the structure and function of the brain through processes such as neuroplasticity and epigenetics, which suggest that a person's experiences and environment can alter their brain's biology.

Heart-Brain Coherence means that the heart and brain work together in a coordinated and complimentary way. Research has shown that the heart sends more signals to the brain than the brain sends to the heart. That comes as a surprise to most of us who always thought the heart was simply an organ that pumped blood through our bodies. Recent studies have shown that the heart can affect our emotions, thoughts, and behaviors through hormones and neurotransmitters in ways we never imagined.

Think of it like this: Heart-Brain Coherence is like a dance between your heart and brain. It's when your heart and brain work together in a cooperative and coordinated way. It's like when you dance with a partner. You look graceful and in sync if you move in the same rhythm. That's what it's like when your heart and brain are in coherence. When your heart and brain are coherent, it makes you feel good and helps your body work better. It can help you feel calmer, happier, and less stressed.

The Institute of HeartMath is a non-profit organization dedicated to advancing the understanding of the intelligence of the heart and its role in physical, emotional, and spiritual well-being. [1] They have done much research on the role of the heart in nurturing a healthy life, not only in physical terms but in psychological and spiritual ways.

Doc Childre, the founder of the Institute of HeartMath, agrees that "the heart is not just a pump. It's an emotional, intuitive, and informative center that communicates with the brain and the rest of the body in ways we are just beginning to understand."

Rollin McCraty, PhD, another leader at HeartMath, said, "The heart is the most powerful generator of electromagnetic energy in the human body, producing the largest rhythmic electromagnetic field of any of the body's organs."

Dan Siegel, MD, clinical professor of psychiatry at the UCLA School of Medicine, said, "The most direct route to higher intelligence is not just through the brain but through the heart." [2]

Joe Dispenza, author, and speaker on mindfulness, neuroscience, and personal transformation, is another leading voice in this area. He said, "When the heart and brain are in coherence, our thoughts become clearer, and our emotions are more stable. We feel a sense of connection and inner peace." [3]

The Biology of Belief by Bruce Lipton is a seminal work in the field of epigenetics and is known for its exploration of the relationship between beliefs, genetics, and health. In the book, Lipton discusses the concept of heart-brain coherence as part of his larger argument about the power of the mind to impact the body.

He wrote, "Heart-Brain Coherence is the cornerstone of a healthy, happy, and fulfilling life. It's a state where the mind and heart are working together in harmony, allowing us to tap into our full potential and live a life full of purpose, connection, and well-being." [4]

Lipton also argues that the heart has a powerful influence on the brain and the rest of the body and that a coherent heart-brain connection can lead to improved health and well-being. He notes that the heart generates a strong electromagnetic field that can impact the brain and other parts of the body. The signals the heart sends to the brain can help regulate the brain's activity, including the release of hormones and neurotransmitters.

Lipton emphasizes the importance of positive emotions, such as love and gratitude, in establishing heart-brain coherence. Heart-Brain Coherence can be measured using tools such as heart rate variability (HRV) analysis, which assesses the variability in the interval between successive heartbeats. High HRV, meaning the heart is able to adapt quickly to changing circumstances, is associated with increased heart-brain coherence. When the heart and brain are in sync, our nervous system is able to respond to stress and challenges in a more balanced and effective way.

. . .

> When thoughts and feelings come together on the same path, moving in the same direction, you will see your faith surge. In other words, when you strongly believe something and also strongly feel it, you now have created a super-highway for manifestation to come rolling into your circumstances in an amazing way.

To use quantum-speak: "Focused intention plus elevated desire manifest quantum leaps." (A quantum leap is an instant jump from one place to another without linear progression.) A biblical way to say the same is, "Genuine faith and sincere prayer leads to miraculous answers."

Elevate Your Faith

We've learned that a sanctified imagination is an important key that unlocks our ability to think the right way - seeing what we desire as if we have already received it. That's the approach Jesus said to take, but what can we do to elevate our emotions? Let's use the word "elevate" as an acrostic to outline some things the Bible and science both teach have the ability to help lift our feelings to align with a faith that works.

E - Exercise regularly.

. . .

Exercise has been shown to have a positive effect on people's emotional well-being by elevating moods and reducing stress and anxiety. This happens because exercise stimulates the release of endorphins, which are naturally occurring chemicals in the brain that promote feelings of happiness and well-being. It can also help improve our self-esteem, leading to increased confidence and positive emotions.

Exercise has been shown to reduce stress by decreasing levels of the stress hormone cortisol and improving mood. So, although you may not instantly see the connection, exercise can bring changes in your body that elevate your feelings and, consequently, position you in a better place to practice a vibrant faith.

> Don't make the mistake of thinking that good spiritual practices have to be religious. Exercise can be a spiritual endeavor. Think of it as "temple maintenance" if it helps. Saint Augustine said, "Take care of your body as if you were going to live forever and take care of your soul as if you were going to die tomorrow."

Bring your heart and mind into coherence by exercising your body which has been proven to release happy hormones that will uplift you.

L – Laugh every chance you get.

Laughing can have a number of positive effects on physical and mental health. Dr. Lee S. Berk is a pioneering medical researcher studying the neuroendocrine and immune effects of positive emotions. He wrote, "Laughter may not be the best medicine, but it certainly is a good one. It has been shown to have physiological and psychological benefits, including reducing stress, boosting mood, and improving immune function." [5]

Quaker theologian and author Elton Trueblood went so far as to say that Jesus had a sense of humor. He wrote, "There is a humor in Christ which was not only refreshing to his contemporaries, but which also can be recaptured by us today. It is not the kind of humor that many of us would anticipate, but it is present in the Gospels, and we miss something of the richness of Christ's personality if we overlook it." [6]

Laughter elevates our emotions, and our emotions fuel our faith. C.S. Lewis said, "Joy is the serious business of heaven." The Bible says there is a time to laugh and that a joyful heart is good medicine. So, at the risk of over-simplifying the topic, laugh your way toward a miracle! Bring your heart and mind into coherence by laughing yourself toward the manifestation you want to see.

E - Embrace mindfulness.

Mindfulness is a form of meditation that involves paying attention to the present moment with openness, curiosity, and acceptance. It is a practice that can help you reduce stress, improve your focus and concentration, and enhance your overall well-being.

By being fully aware of the present moment, we can gain greater insight into our thoughts, feelings, and how we act. It helps us to focus on how we perceive life at any given moment. This goes back to the earlier question, "What do you see?" Being present in the moment gives you a chance to make necessary adjustments and "get your mind right" so that you are in a posture for receiving.

Don't time travel into tomorrow. Stay where "I AM" lives. The Bible says, "Don't worry about tomorrow. It will take care of itself" (Matthew 6:34, Contemporary English Version). Bring your heart and mind into coherence by being present in the moment and refusing to disconnect from what is happening at this exact moment.

Here's a simple plan to guide you in practicing mindfulness:

. . .

1. Find a quiet place to sit or lie comfortably: Finding a peaceful and calm place where you can relax without distractions is important. The goal is to be where you can mentally disconnect from anything that would compete for your attention.

2. Close your eyes and take a few deep breaths to calm your mind and body: Taking deep breaths can help calm you by activating your body's relaxation response, which is the opposite of the "fight or flight" response that is activated during times of stress. It also stimulates the parasympathetic nervous system, which is responsible for slowing down the heart rate, reducing blood pressure, and relaxing the muscles. Knowing these practical benefits for breathing deeply may help you be more intentional about not skipping past it.

3. Bring attention to your breath: Focus on the sense of air entering and leaving your nose or mouth. This will help you anchor your attention in the present moment through intentional awareness.

4. Take note of any thoughts, feelings, or sensations that arise as you focus on your breath without trying to judge them: Acknowledge them without judgment or trying to change them. Instead, accept them and let them pass without getting caught up in them.

5. If your mind starts to wander, gently bring your attention back to your breath: Acknowledge the thought and then let it go. Don't try to push it away or suppress it. Instead, just take note of it with

curiosity and bring your attention back to your breath. Don't judge yourself for having interfering thoughts. Remember that it is normal for thoughts to arise during meditation, and learning to let them go takes practice. Even people who have meditated for a long time have to deal with this.

6. Practice this for 5 to 10 minutes a day, gradually increasing the time as you get more comfortable with mindfulness: Consistency is key when starting a meditation. Repeat this exercise regularly, making it a part of your daily routine: To experience the benefits of mindfulness, it's essential to make it a part of your daily routine.

If you've ever read or heard anything about practicing mindfulness or meditation, you are likely already familiar with these suggestions. The reason they are so commonly found is that they work. Some of us used to think this was only for a bunch of hippies who sat around all day, chanting and talking about their chakras. You know, people who were too out of touch with reality to be of any use to society. Now, many of us know better and understand the value of this practice. Science proves it. The testimony of those who practice it proves it. The Psalmist prayed, "May these words of my mouth and this meditation of my heart be pleasing in your sight, Lord" (Psalm 19:14). Make this a priority and see for yourself.

Remember that mindfulness is a practice that

> takes time to develop. Don't worry about getting it wrong. Be patient and kind to yourself; over time, you will experience the benefits of mindfulness in your daily life.

With regular practice, you will develop a greater awareness of your thoughts, feelings, and sensations, enabling you to respond more effectively to the challenges of daily life. Bring your heart and mind into coherence by embracing mindfulness and learning to live in the moment.

V - Value yourself.

Sometimes the reason we feel drained or discouraged, or even depressed, is that we've given so much of ourselves to other people but have neglected taking care of ourselves. Pointing back to the Law in the Torah, Jesus said that the second greatest commandment is to "love others as you love yourself." How are we to love others if we don't even love ourselves? And how can we find the faith to expect answered prayers if we don't value ourselves?

Learning to love yourself is a huge step toward coming to the place where positive things manifest in your life. If you think little of yourself, you will expect little in life, and you'll get little. You are not undeserving, unworthy, unloved, unsafe, or unappreciated. A parable in Matthew 13 depicts you as a treasure in a field that has such value that Someone gave all He had to purchase you. "You have been bought with a price," and you were worth it.

> You are deserving because you have been chosen by Divine Love. God could have as easily dismissed you as not worth His attention, but instead, He has described you as being His handiwork who is created in His own image and who is beloved. Don't insult what He thinks and has said about you by disputing your own value and refusing to see yourself as you truly are.

If you want your faith to soar, focus on how much you are loved by God and watch your emotions be lifted to align with His heart toward you. And, when that happens, you'll be ready to see mountains move out and miracles move in.

Bring your heart and mind into coherence by valuing yourself.

A – Affirm yourself.

. . .

Affirming yourself can also be a powerful tool for boosting healthy feelings so that your feelings and thoughts will come into alignment. It's important to recognize negative self-talk and replace it with the truth about yourself. To affirm declarations about your own worth is an expression of faith. It is to agree with what God thinks and says about you. It is a biblical perspective, and one scientific research has validated.

The self-affirmation theory is a psychological theory that says that people want to have a positive view of themselves, and when their abilities are questioned, they resist the threat to their self-esteem. Self-affirmations can help them feel better by reminding them of their important values and strengths.

A study published in the *Journal of Social Psychology* found that self-affirmation improved mood and reduced stress in a group of college students.

Take charge of the inner voice and teach it to behave. The messages in your brain have been installed, and you can reformat the programming. When you affirm yourself, you agree with what your Creator has said about your value. You are renewing your mind with truth. You are honoring God and strengthening your faith. Bring your heart and mind into coherence by affirming true thoughts that will enflame your emotions with optimism.

. . .

T - Think supernaturally.

Research has shown that when people have a positive, expansive outlook and a confident attitude, they experience increased motivation and ability, often producing a better outcome. Thinking supernaturally allows you to expand your mind beyond the confines of the physical world and consider possibilities that may not be immediately evident.

> Always remember that your life experiences aren't restricted to what is possible. The Kingdom where you live has a panorama of possibilities that the human mind can't foresee or even imagine. Dare to think supernaturally.

Mother Teresa was a Catholic nun who demonstrated this point. Despite facing continuous hurdles in life, including scarce financial funding, opposition from local authorities, and personal health issues, Mother Teresa imagined something so big it seemed impossible. She founded the Missionaries of Charity in 1950, which grew into a global organization with missions in more than 130 countries. She and her volunteers provided care and support to the sick, homeless, and dying, including those suffering from leprosy and HIV/AIDS.

. . .

Her work often put her at odds with the authorities in India, who saw her as a threat to social order. She faced constant criticism and controversy from some who accused her of being too focused on spiritual matters and neglecting the physical needs of her patients. Despite the critics, she continued to serve the poor and sick, even to the point of often living and working alongside them in the slums.

She received numerous awards and honors during her lifetime, including the Nobel Peace Prize in 1979. Her example continues to inspire people worldwide to push into the places that seem impossible. She demonstrated that when you have a God-given dream in your heart, you can accomplish the unimaginable. She allowed her compassion and commitment to propel her forward until the impossible became possible. Bring your heart and mind into coherence by choosing to think supernaturally instead of naturally.

E - Express gratitude every day.

Practicing gratitude can elevate your emotions by shifting your focus away from negative thoughts and emotions and instead focusing on the positive aspects of your life. When you consciously reflect on what you are grateful for, it helps cultivate a more positive mindset and a greater sense of contentment and fulfillment.

. . .

One way to implement the practice of gratitude is through simple meditations in a way similar to the way we discussed practicing mindfulness. Practicing a meditation on gratitude can be a powerful experience that can profoundly impact your overall outlook on life. By taking time to focus on the things you are grateful for, you can cultivate a more positive mindset, reduce stress and anxiety, and increase feelings of connection and closeness with others.

> When we practice gratitude regularly, we train our minds to focus on the positive aspects of our lives and develop a greater sense of appreciation and contentment.

This can help us navigate life's ups and downs with greater resilience. It can nurture optimism that causes us to really believe that things are possible. Even things we might have thought impossible apart from training our minds through meditation, now become possible to us.

Your spiritual journey may not have included meditation, but many people who have done it would enthusiastically tell you that incorporating gratitude into your life in a structured way can be a simple and effective way to bring positive changes into your life. So, please try this. Take a little while every day to reflect on the things you are grateful for and see how this simple habit can help you.

To show you how this might look in your life, let's practice a meditation on gratitude together for the next few minutes in the following paragraphs. Read this meditation slowly and engage with what you read. Take time to do what you're instructed.

Make the conscious choice to enter into this guided meditation on gratitude. Find a quiet and comfortable place to sit down, become completely relaxed, and take a few deep breaths. Feel tension drain from your body every time you exhale. (Do it before you go further in this meditation.)

Take a moment to focus on your breathing. Allow your breathing to become slow and deep, filling your body with fresh oxygen.

Focus your attention on your heart and imagine it as a beautiful and radiant light. See this light growing stronger each time you inhale and expanding to fill your entire chest with warmth and positivity.

As you continue to breathe deeply, think of the things you are grateful for in your life. This can be anything from your loved ones, to your health, to the simple joys in life.

Let your heart expand with each thing you think of, feeling it fill up with joy and appreciation. Repeat to yourself, "I am grateful for [insert something you are grateful for]," and experience the feelings of gratitude in your heart.

If you find your mind wandering, bring your focus back to your heart and the feeling of gratitude.

Imagine you are surrounded by a beautiful, warm, and comforting light. This light represents all the love, happiness, and blessings in your life. Allow yourself to sink into this light, feeling its warmth embrace you completely.

Allow yourself to stay in this state of gratitude for as long as you like. Take this time to appreciate everything you have and feel the joy and happiness this brings you.

When you are ready, slowly open your eyes and return to your day with a grateful heart, filled with a renewed appreciation for all the beautiful things in your life.

Take a deep breath and smile, feeling the joy and gratitude in your heart. You are now ready to face the day with a positive and grateful attitude.

Remember, you can carry this feeling of gratitude with you throughout the day and into the future. So, whenever you need a boost of positivity, simply stop what you're doing and return to this meditation, allowing yourself to be enveloped by the warmth and light of gratitude.

. . .

These are some simple steps to give you guidance on how to practice heart-brain coherence through the acrostic, ELEVATE – Exercise regularly. Laugh every chance you get. Embrace mindfulness. Value yourself. Affirm yourself. Think supernaturally. Express gratitude daily. They aren't simply inspirational platitudes. They are instructive prescriptions that have proven to work over many years and in many people's lives.

Incorporating these practices that improve heart-brain coherence into your daily lifestyle can have a significant impact on your life. By syncing the heart and brain, you can improve your ability to respond to stress and challenges, increase feelings of calm and happiness, and fortify faith in your life.

I encourage you to take to heart the things discussed here and experience how it will energize your faith when you connect your heart and mind. Don't just pass verdict on what you've read in this chapter. Try it and see how this simple act has the potential to help you bring your feelings and thoughts into agreement in a way that will open the door for the quantum quiff to pop and for you to see how a healthy faith can act as the bridge between what you want and what you get in your life.

Chapter 5

The Quantum World Is the Kingdom of God Manifesting.

Thy kingdom come. Thy will be done on earth as it is in heaven.

–Jesus, in Matthew 6:10

The idea of the Kingdom of God coming into this world has been a central point of discussion in Christian culture since the time of Jesus. To many, it represents the establishment of God's reign on earth and the transformation of society and the world as we know it. The concept suggests that God's presence isn't limited to the spiritual realm but can also be experienced in this world literally and practically. When we pray for the Kingdom of God to come, we are asking for a radical transformation of our world. There are many viewpoints about end times. While they differ in many ways, the common denominator is that each anticipates a time when everyone will clearly recognize the Kingdom of God. This concept challenges us to look beyond our own individual needs and anticipate a world that reflects God's rule of loving grace.

The Kingdom of God Is Here

What does it mean to pray for God's Kingdom to come on earth as it is in heaven? Does the prayer suggest that the Kingdom of God needs to relocate from the celestial realm to an earthly one? Not at all. The prayer isn't about presence but perception.

One time, when the Pharisees asked Jesus when His Kingdom would come, He gave an answer no one expected. He said, "You won't be able to say, 'Here it is!' or 'It's over there!' because the Kingdom of God is already among you" (Luke 17:21). Their question was rooted in an underlying sense of nationalism that only understood physical kingdoms built around political parties and powerful people who rule over others who submit. They wanted a map of things to come, but Jesus told them that what they were looking for was already here. They just couldn't see it. The Kingdom of God isn't out there but is already among us.

Many in the Evangelical world today have the same problem as those did two millennia ago. They want to know a timeline for when Jesus Christ plans to set up His Kingdom, and, like the Pharisees, they can only anticipate a future time. They imagine Him coming back, taking over, and setting up a kingdom on the earth that is absent at this point in time. In their minds, we have little chance for things to get any better until Christ returns and takes over by force. Remember the earlier example about the greyhounds who chase the mechanical rabbit around the racetrack but can never catch it? That's the kind of approach many take to the Kingdom of God. They're chasing after something they think is out there, ahead of us, but in reality, is already here among us, and, in fact, within us.

> While Christ is indeed coming again, in their obsession with future things, many people miss the point that He is here right now.

Whatever your views may be about end-times' prophecies, it's important to acknowledge that Christ is here, and, according to Him, the Kingdom of God is already here, too. We will not be able to see the Kingdom of God manifesting among us as long as we believe it's only a future event.

This Kingdom Jesus described isn't one like anything we've known so far. The kind of Kingdom He described is characterized by compassion, mercy, and forgiveness. It is a Kingdom open to anyone and everyone willing to enter by faith. It is a life of service toward others and one of humility and selflessness.

. . .

The Kingdom of God coming on earth is nothing less than our eyes being opened to the reality that already exists among us. God's Kingdom won't be manufactured in this world but will be manifested. Remember that faith doesn't create but facilitates what is already present. Faith simply brings outcomes from the invisible to the visible.

The Kingdom of God will become visible to us all, but not because the King and His Kingdom aren't already here. "Thy kingdom come" isn't to be tied to our calendar but to our consciousness. While His Kingdom is already among us, most aren't conscious of it because it isn't the kind of kingdom we've been taught to recognize. Many are waiting for something to happen when, in reality, "it is finished."

In C.S. Lewis's last work of fiction, *Till We Have Faces*, he tells the story of two sisters, Orual and Psyche, princesses in the Kingdom of Glome. Everything is going well until the Priest of the goddess Ungit comes to the King to tell him that Psyche must be sacrificed to the goddess. Psyche is drugged and chained to the sacred tree, where she is left to be eaten by the shadow brute.

A few days later, Orual returns to the tree to give her sister's bones a proper burial. When she arrives, Psyche's bones are nowhere to be found. She wanders to the river, crying when she sees Psyche standing on the other side. Orual is shocked. She doesn't know what to think. How is it possible? She knows that Psyche is dead. How can this be true?

Orual crosses the river, and she and Psyche embrace. Psyche then tells her sister the story of how the God of the west wind saved her from the shadow brute and brought her to his palace to be his bride. Orual thinks Psyche has lost her mind. But, to humor her, she listens to her sister's story as if she believes it.

Psyche leads Orual a short distance away to sit in the heather. There she serves a glass of wine to Orual, the choicest wine in an exquisite goblet. She asks her if she likes the goblet and the wine. Orual goes along with her and nods, but what she actually sees is her sister cupping her hands in a pool of water. She becomes increasingly convinced Psyche has lost her mind.

Psyche goes on to tell Orual stories of gods and palaces and how she wears the most beautiful gowns. Orual sees no palace, only woods; no gowns, only Psyche dressed in rags. After a while, she can bear it no longer and demands that her sister show her the palace.

Orual is dumbfounded when her sister nods with a smile and says, "Of course, I will. Let us go in."

Orual asks, "Is it far?"

"Far to where?" Psyche responds.

"To the palace," Orual shouts, "to your god's house!"

Psyche starts to tremble. "Orual, what do you mean, is it far?"

"Mean?" Orual asks.

"Where is the palace? How far have we to go to reach it?"

Psyche starts to cry. Through her tears and cries, she stares into Orual's eyes and answers, "But this is it, Orual! Can't you see it? You are standing on the stairs to the great gate!"

This kingdom required clear perception for there to be conscious participation. So it is with the Kingdom of God. The King we await is already here, and His Kingdom exists among us as surely as it exists beyond us. "Thy Kingdom come. Thy will be done on earth as it is in heaven." The Kingdom simultaneously exists on earth and in heaven. Jesus told His disciples that they weren't able to recognize the Kingdom of God among them because they were looking for it in the wrong places.

> People today are trying to establish the Kingdom of God by having the right religious programs, practices, politicians, and other misguided approaches that fail to recognize that we don't need to create God's Kingdom but only need to connect to it by faith.

. . .

In Mark 1:15, Jesus told His disciples, "The time is fulfilled, and the Kingdom of God is at hand." When He said the Kingdom is at hand, did Jesus mean the time was close or that the Kingdom of God was nearby? Was He speaking of a date or distance?

In the original language of the New Testament, the phrase "at hand" is *ēngiken*. Another time the word was used was at the last supper Jesus had with His disciples. Speaking of Judas, the one who would betray Him, Jesus said, "the one who betrays me is at hand" (Matthew 26:46). Was that a reference to time or location? Obviously, He meant that the one standing right there, nearby, was the one who would betray Him. So, when the Bible says that the Kingdom of God is at hand, it refers to its proximity to us.

Another time, in Matthew 4:17, the Bible says, "From that time on Jesus began to preach, 'Repent, for the kingdom of heaven has come near.'" We need to stop thinking that the Kingdom of God isn't here or even that it isn't already fully here.

Some people speak of "extending the Kingdom of God," but that terminology can be misleading. While it recognizes the presence of God's Kingdom among us, it fails to acknowledge its omnipresence. The manifestation of the Kingdom of God will come through a spiritual awakening to its pervasive presence. As that happens in our lives, we will find ourselves proclaiming the same message Jesus did: "Change your mind about what you've expected! The Kingdom of God is right here!"

Over a hundred and fifty years ago, H. Ernest Nichol grasped this truth and wrote a hymn that celebrated it:

"We've a story to tell to the nations,
That shall turn their hearts to the right.
A story of truth and mercy,
A story of peace and light.

For the darkness shall turn to dawning,
And the dawning to noonday bright,
And Christ's great come Kingdom shall come on earth,
The Kingdom of love and light."

When it comes to empowering your faith, nothing is better than realizing that all you desire is already here with you and that it's only a matter of seeing it so you can seize it. Thankfully, there is a quantum reality that helps to explain this spiritual truth.

WAVE-PARTICLE DUALITY

Wave-particle duality is a fundamental concept in quantum mechanics that describes how particles can exhibit both wave-like and particle-like behavior. Is it a wave, or is it a particle? Until it is observed, the answer is, "Yes, it is both." It makes no sense, even to the scientists. How can something exist in two ways and places at the same time? It's baffling. Matters of spirit usually are.

Richard Feynman, one of the most renowned physicists of the 20th century, once said, "I think I can safely say that no one understands quantum mechanics. Do not keep saying to yourself, if you can possibly avoid it, 'But how can it be like that?' because you will get 'down the drain' into a blind alley from which no one has yet escaped. No one knows how it can be like that."[1]

Is the Kingdom of God in heaven or on earth? The answer is mindboggling, but the answer is "yes." "My Kingdom is not of this world," said Jesus in John 18:36, and then He turned right around and said, "The Kingdom of God is among you" (Luke 17:21). Wait. Which is it? Is God's Kingdom not of this world, or is it among us in this world? The answer is both.

> All the possibilities for your life exist inside the Kingdom of God and are available to those who see. It is when we see through the eyes of faith that these possibilities manifest in our lives in definitive ways. The Kingdom, the King, and His gifts are all fully accessible right now. It's simply a question of perception. A vision of the eternal reality that envelops your earthly experience changes everything.

. . .

Wave-particle duality suggests that the nature of reality is not set but is malleable. Heaven on earth is available to you as you walk in faith. Your life will change when your temporal perspective aligns with the eternal truth of the Kingdom of Kindness ruled by the King of Kings, whose desire is to do "exceedingly, abundantly above all that you ask or think" (Ephesians 3:20). "Thy Kingdom come. Thy will be done on earth as it is in heaven" is the experience you are intended to know and can know as your faith rises and resonates with the reality that surrounds you.

One school of thought about the Kingdom of God suggests a "now and not yet" aspect that recognizes the Kingdom is here but not fully here yet. Those who embrace this position point to how Jesus used the image of yeast to communicate that the Kingdom is small now but will expand like yeast does for bread.

What if there is a better way to understand it? What if the expansion illustrated in that metaphor isn't about the presence of the Kingdom but rather about its perception?

Is Jesus Christ fully here with us in this moment? "Not physically," one may suggest, but a question begs to be answered: "Is the main element of the Kingdom of God its physicality?" What is the characteristic that defines the Kingdom of God? Is it the material manifestation or the spiritual manifestation of that Kingdom? Yes, believers anticipate the coming of Christ, but must we wait until then to fully enjoy the benefits of His Kingdom?

During his trial, Jesus was speaking to Pilate, the Roman governor. Pilate asked Jesus if He is the King of the Jews, and Jesus responded by saying, "'My kingdom is not of this world. If it were, my servants would fight to prevent my arrest by the Jews. But now my kingdom is from another place" (John 18:36).

> Jesus clarified to Pilate that His Kingdom is not a physical, political kingdom like the Roman Empire but a spiritual one that transcends the physical world. It is a Kingdom not limited by geographical boundaries or political structures.

If you don't want to wait until you die to experience the reality of living inside the Kingdom of God, recognize that you are already there. At this very moment, you are already "seated in heavenly places" (Ephesians 2:6) and possess everything you need to live at the highest level. So you don't have to die first. British theologian and scholar N.T. Wright said, "The kingdom of God is not a matter of getting people to heaven, but of getting heaven to people."

. . .

"Thy Kingdom come. Thy will be done on earth as it is in heaven" is a prayer for manifestation. By aligning yourself with the culture of that Kingdom of Christ, you can tap into infinite potential and know heaven on earth. Jesus said, "God has blessed you because your eyes can see and your ears can hear" (Matthew 13:16, Contemporary English Version). So open your eyes now and see. That's where you are in your journey at this point.

Wave-particle duality is the scientific expression of how energy can simultaneously be an all-encompassing wave and a specific particle. So it is with the Kingdom of God, which is at the same time everywhere and right here with you in your particular circumstances.

QUANTUM ENTANGLEMENT

Quantum entanglement refers to the phenomenon in which two or more particles are connected in a way that the state of one particle can't be described independently of the other particle, even if the particles are separated by a large distance. This means that if the state of one particle is changed, the same thing will instantaneously happen to the other.

In terms of being simultaneously in two places, quantum entanglement suggests that particles can exist in multiple states or locations at the same time, a concept we identified earlier in this book as "superposition." The idea of a particle being in two places at once is a result of wave-particle duality.

. . .

Waveform and particle form are both present until one particle is observed. Then, the superposition collapses at that point, and the particle manifests in only one state. This idea of particles being simultaneously in two places was described by Einstein as "spooky action at a distance" and continues to be a mystery to the scientific understanding of the world and the nature of reality.

This quantum phenomenon offers a framework for better understanding how faith can cause your prayers to be answered. Using the language of science, "the superposition collapses, and the particle is then in only one state or location." Using a biblical perspective, it is by seeing through the natural into the supernatural that faith-empowered observation brings the answer to our prayers. As discussed in an earlier chapter, "what you see is what you get."

The eternal reality that envelops us becomes an experiential reality that encourages us to live in faith as if we "have been given everything we need for life" (2 Peter 1:3) because we have. It's a different way of looking at life that is more than hope. It is a confidence that comes from understanding that what is ultimately true in the supernatural world is unambiguously true in the natural world. Quaker author, philosopher, and educator Elton Trueblood rightly said, "Faith is not belief without proof, but trust without reservation." [2]

> This isn't a mind game. It's real, and we experience it when we can look beyond our temporal walk into the transcendent world that defines the ultimate reality in our lives. It's not a religious "fake it 'til you make it" mantra we mindlessly recite, as some groups that focus on faith would have us practice. Instead, this is a substantive, scientific, and spiritual reality that empowers us to faith it, not fake it.

In my book, *Quantum Life*, I discussed in detail the power in our spoken words: "Jesus said, 'By your words, you will be justified and by your words you will be condemned'" (Matthew 12:36) Justified or condemned by whom? By God? No! By life itself! The words we speak are a prophecy of what our lives will be. Your words stir energy into motion to facilitate the thing you say. We are not victims who watch life happen to us unless that's what you want to be. According to your faith, it will be unto you, but don't create a personal storm and then complain when you get wet. And the horrible thing is that some people keep complaining about the rain and never come out of it!

The man known as the wisest of all men said, "The one who guards his mouth and tongue keeps himself out of trouble" (Proverbs 21:23) and "Wholesome speech is a tree of life" (Proverbs 15:4). Don't underestimate the power of the words you speak. Every one of them has a frequency, so make sure that your words are in resonance with the frequency of the Kingdom of God.

If you heat a plate of food in the microwave oven, the meal becomes hot because of frequency. The frequency of a microwave oven is typically 2.45 GHz (gigahertz), equivalent to a wavelength of approximately 12.2 centimeters. The microwave oven contains a magnetron, a device that converts electricity into high-powered microwave radiation. When you turn on the microwave, the magnetron produces these microwaves, which pass through the microwave oven and into the food.

Water molecules, fats, and sugars in the food absorb microwaves. The higher-frequency microwaves interact with these molecules and cause them to vibrate rapidly, which generates heat. This heat then spreads throughout the food. The science of this process gives proof that frequency affects matter. Energy changes physical things.

Since your words are an energetic frequency, what you say has a literal ability to change things. In the beginning, God said, "Let there be," and matter manifested by His word. "What is seen was not made out of things that are visible" is how the Bible describes it in Hebrews 11:3. He "calls into being things that do not yet (physically) exist," says Romans 4:17. How exciting it is to understand that process better today! What used to require blind faith based only on Scripture now has become a bold faith based on

Scripture and science. While some would argue that Scripture needs no scientific proof, it certainly empowers our faith to now understand the "how" behind the "what" the Bible tells us.

What is the message of faith taught in the Bible? Paul said in Romans 10:8, "The word is near you; it is in your mouth and in your heart, that is, the message concerning faith that we proclaim."

> The message he taught about faith was that you find it in your mouth. Speak the language of the Kingdom of God, and you'll see the manifestation of that Kingdom. Created in His image, you possess a creative power expressed through your words.

This kind of discussion might trigger negative associations with some readers. It may remind them of fringe groups or extreme sects who have taught similar things from a skewed perspective. Don't forfeit the importance of this truth because of a misguided judgment of guilt by association.

New Agers have spoken of this reality but leave Christ out of the equation. That is a fatal flaw because, as King, He is the source and center of the Quantum Kingdom discussed here.

. . .

Some religious groups have stressed the "word of faith" as the key but reduced the message to religious magic that, not unlike the New Age Movement, usurps the centrality of Christ with formulas that supposedly bind and obligate God to do what we want. Not everything cloaked in religious language is Christ-centered.

Some have hijacked this Kingdom principle for their own personal enrichment. We've all heard of the prosperity preachers who have culturally appropriated this aspect of the Kingdom of God for their own selfish reasons. Countless self-help books have been written to explain how this approach can make you rich and successful.

Those self-serving approaches are wrong. They are dead wrong, but the abuse doesn't nullify the truth. A healthy verbal expression of quantum faith brings us into cooperation with the Spirit so that we effortlessly but intentionally benefit from the supernatural laws of the Kingdom. Don't let someone who has misappropriated this scientific and biblical practice keep you from appropriating it in your own life.

It's a sad reality that this kind of teaching needs caveats because of the abuse that exists inside Christendom as well as outside. Don't let those who have misused this powerful truth turn you away from this important key to unlocking the privileges of your citizenship in heaven. You live in union with Christ and have all the advantages of the quantum entanglement that defines your connection to Him. Christ is not in your life; He is your life. "In Him, we live and move and have our being" (Acts 17:28).

This quantum entanglement that we share with Him can be seen in the Apostle Paul's words, "For me to live is Christ" (Philippians 1:21). Colossians 3:4 is another clear case for our entanglement: "When Christ, *who is your life*, appears, then you also will appear with him in glory" (emphasis added). Christ isn't simply *in* your life. He *is* your life. Entanglement suggests that what is true of Him is true of you. In fact, the Apostle Peter said that you are "a sharer of the divine nature" (2 Peter 1:4).

Faith Is a Verb

So, if you and Christ share a connection to the extent that quantum entanglement and biblical teaching demonstrate, it's simply a matter of living as if it's true. Christ defines you, so the only thing to do to see His Kingdom come and His will be done in your daily lifestyle is to recognize your union with Him and look to Him to animate your actions.

Living at your highest level inside the Kingdom of God means understanding that this divine quantum entanglement will energize your faith to act. If you were to ask the average person what form of speech the word "faith" is, the answer would most likely be "a noun." While that wouldn't be incorrect, many don't know that the word is also a verb. The *Oxford English Dictionary* explains,

"The verb faith is not very common, and is limited chiefly to expressions in which it represents the practice or habit of putting trust or confidence in someone or something." [3]

The Bible teaches that faith, apart from action, is nothing more than a lifeless idea. "As the body without the spirit is dead, so faith without deeds is dead" (James 2:26 NIV). Faith is our active participation in manifesting all God has planned for us.

Quantum faith is an energetic faith. It's not a passive lifestyle. St. Ignatius of Loyola said, "In God's eyes, our words have only the value of our actions." In other words, "talk is cheap." Our life here is a grace *walk*, not one of passivity. While there are times when waiting is appropriate, there are other times when Spirit-animated action expresses our faith.

> Living daily with the knowledge that our lives are entwined with the Kingdom of God can be a powerful source for empowering our faith. As we recognize the entanglement we share with God, we are able to tap into a source of strength and wisdom that transcends our human limitations.

The Kingdom of God is not out there but is right here at this very moment. His Kingdom comes, and His will is done in your life as it is in heaven when you recognize its present reality and behave accordingly. Wave-particle duality shows both the transcendent and imminent reality of God's Kingdom, and quantum reality shows that Christ in, through, and as you can express a lifestyle that is nothing less than miraculous.

Conclusion

Coming to the end of this discussion about these quantum realities that can empower your faith, it's important to understand how the quantum science explained in this book is intended to be applied. The most important thing to know is that these quantum truths help describe the culture of the Kingdom of God. They give insight into how the Creator has set particular protocols in place that allow us to participate in His creation in a way that maximizes the opportunity to live life at the highest level.

On the other hand, it is important to understand that the Kingdom of God can't be reduced to quantum elements consisting of subatomic particles. If you have thought otherwise, I've missed the mark. It must be about the King before it's about the Kingdom. I've assumed you who have journeyed with me through the years don't need me to start at the kindergarten level every time I write another book. If this is your first introduction to my writing, I encourage you to read some of my earlier books for

Gospel 101. The science discussed here simply offers a deeper understanding of how God has designed the world to work. As followers of Jesus Christ our lives are not rooted in principles, no matter how helpful they may be, but in a Person. With that understanding, this book is intended to be a travel guide offering practical direction for your journey.

Modern science is not God but certainly reveals God. This has always been true. For instance, cosmology is a language through which God has always spoken. Romans 1:20 says, "For since the creation of the world God's invisible qualities—his eternal power and divine nature—have been clearly seen, being understood from what has been made, so that people are without excuse."

Astronomy is another language the Creator speaks to His creation. Psalm 19:1 teaches, "The heavens declare the glory of God; the skies proclaim the work of his hands." Stars and planets aren't God, but He certainly makes Himself known through them.

Your mother tongue may be the language of faith but not everyone speaks that language. The King of Kindness, who is the source and center of the Kingdom of God, can speak whatever language a person understands. Job 12:7-8 says, "But ask the animals, and they will teach you, or the birds in the sky, and they will tell you; or speak to the earth, and it will teach you, or let the fish in the sea inform you."

So, don't mistakenly conclude that the quantum realities discussed here seek to reduce Christ to scientific elements. To think that is to miss the point altogether. Understanding the five scientific realities discussed in these chapters is akin to understanding the practical instructions given in the epistles of the New Testament. The instructions aren't to detract from Christ as your source but rather are intended to show how to live the Christ life. They are a road map to help guide you in your walk inside the Kingdom of God as you are animated by the Spirit of Christ.

The man who wrote more than half the New Testament, the Apostle Paul, wrote, "The life which I now live in the flesh I live by the faith of the Son of God" (Galatians 2:20, KJV). The faith of Jesus Christ is yours. It's simply a matter of knowing how to practice that faith. When you know that and do that, you will experience the Kingdom of God now. It's not something you have to wait until you die to experience.

Quantum principles are not only the building blocks of our physical world, but they also have the power to transform our personal lives. By consistently applying these principles, we can learn to recognize the interconnectedness of all things and tap into our divine potential. Let us not be afraid to take the quantum leap and embark on a journey of self-discovery and growth. With dedication and practice, we can unlock new levels of spiritual awareness and a higher level of life. Embrace the power of the quantum world and allow it to enrich your life in ways beyond your wildest imagination.

Afterword

This book is a result of the way my own journey has progressed over the past years. It started when I began to read books on quantum mechanics as a hobby. I was immediately struck by the mystical elements of this nascent science that stood in sharp contrast to the scientific materialist's approach we were all taught in school.

As I continued to study, I began to see what looked like similarities between truths inside quantum mechanics and the biblical teaching I had always known. As time progressed, I saw the two come together to the extent that I finally recognized that they aren't just similar. Modern, quantum science is teaching us some of the same things as the Bible, but in a different language. Once I saw that, I couldn't get enough of it. It is thrilling to see the topic of faith clarified in the language of science as I've described in this book.

If you want to know more, you may be interested in an online subscription group I teach weekly. In that community, we discuss practical ways to express grace in many areas of life. The teachings are available to watch at the viewers' convenience. It's a great group of people from diverse walks of life. We would be happy to have you join us. You can get information about this community at www.gracewalkexperience.com. Additionally, you can find more complete information about all I do at www.stevemcvey.com.

Thank you for investing your time in this book. May the Spirit of Truth continue to guide you into an ever-expanding knowledge of Life.

Endnotes

Introduction

1. https://www.cnbc.com/2018/02/01/google-ceo-sundar-pichai-ai-is-more-important-than-fire-electricity.html
2. https://quotes.thefamouspeople.com/werner-heisenberg-5202.php
3. https://www.goodreads.com/author/quotes/5569343.Freeman_Dyson?page=3
4. John Archibald Wheeler, "At Home in the Universe," in *Cosmic Frontiers: New Discoveries in Astrophysics*, edited by Richard A. Muller (W. H. Freeman, 1981), p. 8.
5. Carl Sagan, *The Demon-Haunted World: Science as a Candle in the Dark*, (Ballantine Books, 1997), p. 29.

CHAPTER 1

1. Taken from Yale News at https://news.yale.edu/2015/12/07/negative-beliefs-about-aging-predict-alzheimer-s-disease-yale-led-study
2. https://www.sciencedirect.com/science/article/pii/S1469029222001522
3. https://journals.lww.com/psychosomaticmedicine/Abstract/2019/10000/Stress_and_Illness__A_Role_-for_Specific_Emotions.7.aspx
4. https://journals.sagepub.com/doi/abs/10.1177/01461672211072823
5. Taken from a paper published by Pascual Jordan in the journal *Zeitschrift für Physik* in 1927, which laid out some of the fundamental principles of quantum mechanics.

CHAPTER 2

1. Lipton, Bruce. *The Biology of Belief: Unleashing the Power of Consciousness, Matter & Miracles.* Carlsbad, CA, Hay House, 2005, p. 147.
2. The late William James is widely regarded as one of the most influential thinkers in the development of psychology and philosophy in the late 19th and early 20th centuries. He was a professor of philosophy at Harvard University and was one of the first to offer a psychology course in the United States. James is best known for his work on the principles of psychology,

including his landmark book *The Principles of Psychology.*

3. The source of this quote is an essay titled "The World As I See It," taken from a book by Albert Einstein called *Out of My Later Years.*
4. Martin Luther King Jr., *Strength to Love* (Minneapolis: Fortress Press, 2010), 96.
5. Fritjof Capra is an Austrian-born physicist, systems theorist, and author who has written extensively on the intersection of science, philosophy, and spirituality.
6. Taken from Frank Herbert's science fiction novel "Dune," which was first published in 1965. The quote is from a character named Duke Leto Atreides, one of the main characters in the novel. In the story, the character uses the quote as a metaphor for the need to be aware of and adapt to changing circumstances.
7. Francis Collins is an American physician and geneticist who is known for his leadership in the Human Genome Project, an international research effort that mapped and sequenced the entire human genome. He is also a devout Christian and has written extensively on the relationship between science and faith.

CHAPTER 3

1. https://www.jpsmjournal.com/article/S0885-3924(12)00094-2/fulltext

2. McVey, Steve. *Quantum Life: The Kingdom of God is Within You*. The Writer's Society. 2023.
3. https://www.ncbi.nlm.nih.gov/pmc/articles/PMC7151200/

CHAPTER 4

1. For further study on heart-brain coherence, the HeartMath institute is a good resource and can be accessed at www.heartmath.com.
2. https://drdansiegel.com
3. Joe Dispenza offers a plethora of content to help in this area. You can learn more about him and his teaching at https://drjoedispenza.com
4. Lipton, Bruce H. *The Biology of Belief: Unleashing the Power of Consciousness, Matter & Miracles*. Carlsbad, CA: Hay House, Inc., 2005.
5. http://www.dr-lee-berk.com/real-health-benefits/#.ZGVAqS-B1pQ
6. This quote comes from Elton Trueblood's book *The Humor of Christ*. The book has been widely quoted and referenced by many scholars and theologians in discussions about the role of humor in Christianity.

CHAPTER 5

1. Richard Feynman, "The Great Conservation Principles", in *The Character of Physical Law* (Cambridge: The MIT Press, 1994), 129.

2. D. Elton Trueblood Quotes. BrainyQuote.com, BrainyMedia Inc, 2023. https://www.brainyquote.com/quotes/d_elton_trueblood_130096, accessed May 17, 2023.
3. *Oxford English Dictionary*, s.v. "faith," n. and v.) I cite an authoritative dictionary here because I want you to see that faith is an action word.

Endorsements

WHAT OTHERS HAVE SAID ABOUT STEVE MCVEY'S BOOKS:

Grammy Award Winning Gospel Singer, Kirk Franklin said, "Steve McVey's books have been used by God to transform my Christian walk."

Dr. Tony Evans, President of The Urban Alternative wrote, "My good friend, Steve McVey, has put the amazing back into grace."

Gary Smalley, author of The Language of Love: "Few people have had the life change effect on my life that Steve McVey has had. Whenever I hear that he has a new book, I buy several copies."

Neil Anderson, author of The Bondage Breaker: "(Steve

McVey's book) A Divine Invitation will enlarge your heart and increase your comprehension of God's love that goes beyond knowledge."

The late Bill Bright, Founder of Campus Crusade for Christ, wrote about A Divine Invitation: "Steve McVey has given us in very clear and understandable language a wonderful, indelible picture of just how beautiful, complete and even startling God's love for us really is."

Fourteen Grammy Award winner, Ron Block
"Steve McVey is one of the voices unafraid to tell the whole truth about grace. Steve's biblical, solid, and life-changing writing points me to the freely-given love, favor, and grace of God in Christ - a grace walk experience.'

ABOUT THE AUTHOR

Dr. Steve McVey is the author of nineteen books, including the best seller, *Grace Walk*. Steve writes to address specific needs in the reader's life. His books are filled with biblical truth, practical application, humor and and affirmation that will encourage you and strengthen you in your own journey of faith.

Steve and his wife, Melanie, live in the Tampa Bay area of Florida. They have four adult children and five grandchildren.

Also by Steve McVey

Grace Walk

The Secret of Grace

Grace Amazing

52 Lies Heard in Church Every Sunday

Grace Walk Devotional

Helping Other Overcome Addictions (with Mike Quarles)

Unlock Your Bible

A Divine Invitation

Walking in the Will of God

When Wives Walk in Grace

The Godward Gaze

Grace Walk Moments

Getting Past the Hurt (with Melanie McVey)

Beyond an Angry God

The Grace Walk Experience Workbook

Journey into Intimacy Workbook

Anchored: Five Keys to a Secure Faith

Quantum Life

Publisher Information

QUANTUM FAITH was published by The Writer's Society a Grace & Trinitarian Publishing Company empowering authors to release books that reveal God's goodness to the world through a variety of self publishing services.

www.thewriterssociety.online

Made in the USA
Las Vegas, NV
15 October 2023

79101034R00072